Jessica Piroli
Marcia Xavier Peiter
Adroaldo Dias Robaina

Cutting Gerbera Irrigation

Jessica Piroli
Marcia Xavier Peiter
Adroaldo Dias Robaina

Cutting Gerbera Irrigation

Productive aspects and economic return

ScienciaScripts

Imprint
Any brand names and product names mentioned in this book are subject to trademark, brand or patent protection and are trademarks or registered trademarks of their respective holders. The use of brand names, product names, common names, trade names, product descriptions etc. even without a particular marking in this work is in no way to be construed to mean that such names may be regarded as unrestricted in respect of trademark and brand protection legislation and could thus be used by anyone.

Cover image: www.ingimage.com

This book is a translation from the original published under ISBN 978-613-9-61353-3.

Publisher:
Sciencia Scripts
is a trademark of
Dodo Books Indian Ocean Ltd. and OmniScriptum S.R.L publishing group

120 High Road, East Finchley, London, N2 9ED, United Kingdom
Str. Armeneasca 28/1, office 1, Chisinau MD-2012, Republic of Moldova, Europe
Printed at: see last page
ISBN: 978-620-7-63873-4

Table of contents:

CUT GERBERA IRRIGATION
Productive aspects and economic return

Jéssica Dariane Piroli
Marcia Xavier Peiter
Adroaldo Dias Robaina

Santa Maria - RS

SUMMARY

IRRIGATION OF GROWING EGGS

Productive aspects and economic return

Gerbera is traditionally grown as a cut flower and is one of the most commercially important ornamental plants in Brazil. However, this activity has been undergoing transformations, making the production chain increasingly competitive and requiring decision-making skills, including irrigation management in a protected environment, where production depends almost exclusively on irrigation water. With this in mind, the aim of this study was to determine the effects of different limits of water availability on the cultivation of the Caribá cultivar of cut gerbera, evaluating the production of floral stems and identifying the appropriate management for the production of the crop in pots with substrate. The experiment was conducted in an air-conditioned greenhouse in the Floriculture Department of the Polytechnic College of the UFSM, Santa Maria - RS. The experimental design used was entirely randomized, with five treatments and ten replicates, making a total of 50 experimental units. Five limits of water availability were evaluated, corresponding to 120%, 100%, 80%, 60% and 40% of the pot's holding capacity. Pots with a capacity of 25 liters were used and one seedling was transplanted per pot, i.e. each pot was an experimental unit, making a total of 50 pots. The crop coefficient (kc) for cut gerbera was determined and the following variables were analyzed: stem length (cm); inflorescence diameter (cm); stem diameter (mm) and number of stems per plant. The results showed that the crop coefficient for gerbera, grown in a protected environment, was 0.72 for vegetative period I, 0.81 for vegetative period II, 0.85 for flowering period I, 0.89 for full flowering period II, 1.33 for full flowering period III and 1.06 for full flowering period IV. It was found that water availability of 80% and 100% is suitable for producing long stems with diameters of between 5 and 6 mm, which has a direct impact on the quality of the final product. The maximum technical efficiency for the production of gerbera stems was observed with a water table of 79.3% of the pot's holding capacity. The lamina that corresponds to the maximum economic efficiency was 70.6 mm, and the profit for different types of stems increases when the ratio of the water factor to the product is lower.

Keywords: Cut flower, irrigation management, production components.

Chapter 1

1. INTRODUCTION

Cut flowers are one of the most important products of floriculture, and due to their fragility, these species require care throughout the production process. Among the main species of cut flowers grown in Brazil are roses (30%), chrysanthemums (15%), lisianthus (12%), lilies (7%) and gerberas (6%). The main destination for the production of flowers and ornamental plants in Brazil is the domestic market, and the sector is expected to grow by 9% with revenues of R$7.2 billion (IBRAFLOR, 2017).

Considered one of the five most important cut flowers on the domestic market, *gerbera* (*Gerbera jamesonii*) is renowned for the diversity of its cultivars, especially its wide variety of colors and shapes. It is widely used, especially in arrangements and bouquets (GUERREIRO et al., 2012; PATRA; BEURA, 2016).

Rio Grande do Sul is the third largest consumer of flowers and ornamental plants, with consumption of more than 38 reais per capita, and gerbera is the third best-selling flower, surpassed only by roses and chrysanthemums (LIMA JÚNIOR et al., 2015; SEBRAE, 2003). In the central region of the state, in the municipality of Santa Maria/RS, the best-selling cut flowers are roses and gerberas (MENEGAES, 2015).

The gerbera crop is growing all over the country, generating income and jobs in the producing regions. In order to make a profit, it is important to analyze a number of relevant factors in relation to the quality of the plant, including correct irrigation management. The correct management of irrigation in cut flowers affects the final quality of the product, as well as its productive potential (MENEGAES, 2017), and implies greater efficiency in the use of water, especially in protected cultivation where considerable results are found regarding productivity parameters (PANJ et al., 2014).

The protected cultivation system is ideal because of the adverse weather conditions. While open-air crops receive effective rainfall and irrigation is complementary to satisfy the crops' water needs, in greenhouses, the only source of water is provided by irrigation (SOARES et al., 2010; GIRARDI et al., 2017).

Therefore, water is one of the most important factors to consider in order to maximize production and improve the quality of flower stems. In this sense, verifying the response of ornamental plants subjected to different levels of soil moisture requires studies into correct water management. It is known that deficient irrigation has a direct impact on yield reduction, as it only replenishes water in the surface layers and does not reach the roots, while excessive irrigation causes nutrients to leach from the substrate and favors the proliferation of pathogens, thus damaging the quality of the flowers (SOARES et al., 2010; GIRARDI et al., 2012).

In the flower production system, whether in pots or beds, the substrate has been used in most commercial crops of ornamental species, including gerbera. The substrate used has a direct influence on the growth of plants grown in containers, as the root system is reduced to the space available in the pot, limiting root growth and the amount of water stored, unlike when grown in soil (FERMINO, 2002; LUDWIG et al, 2015).

Due to the growing demand for cut flowers, there is currently a need for information on management, especially regarding irrigation in protected cultivation, which is directly related to quality and visual appearance, considering the climatic variations in Rio Grande do Sul. Thus, studies in this direction help to improve the efficiency of the highly technified production process, with a view to correct management, which will increase crop yields (VEATCH-BLOHM et al., 2012; FAVA; CAMILI, 2014).

Numerous research studies involving factors such as water depth and product aim for maximum physical productivity, without considering the economic aspect, a factor that should be analyzed, as the economic optimum does not always correspond to maximum physical productivity (VALERIANO et al., 2017). Water is one of the most important factors to consider when growing flowers, as its benefits include increasing yields and improving the aesthetic quality of flower stems.

The aim of this study was to determine the effects of different water availability limits on the production of cut gerbera flower stems grown in pots with substrate.

Chapter 2

2. LITERATURE REVIEW

The aim of this literature review is to provide a theoretical basis for the topic proposed in this work. Thus, this chapter presents a literature review on the *gerbera* (*Gerbera jamesonii*) crop, through a brief history on the general aspects of the crop, production and its ornamental relevance, as well as the use of substrates for the production of cut flowers and the importance of irrigation for crop management, especially in a protected environment. Finally, we present existing research on this subject.

2.1 GERBERA CULTIVATION

Gerbera jamesonii, commonly known as gerbera, is native to southern Africa and Asia. It is a herbaceous ornamental species with a chapter inflorescence belonging to the Asteraceae family, as well as the *chrysanthemum* (*Chrysanthemum* ssp.) and the sunflower (*Helianthus annuus* L.) (SOUZA; LORENZI, 2008). It stands out for the diversity of colors and shapes of its inflorescences (ROGERS; TJIA, 1990; ROMAHN, 2007). Its cultivation, as well as the entire market for flowers and ornamental plants, is in full expansion phase (HULSHOF, 2008).

The gerbera is known in Brazil as a cut flower, being among the five most commercialized flowers, its use is wide, mainly in the composition of arrangements and bouquets (GUERREIRO et al., 2012; PEREIRA, 2013; PATRA; BEURA, 2016).

Under natural growing conditions, the gerbera has a root system that is initially pivotal and becomes fasciculated as it develops, a subterranean stem and rosette-shaped leaves, which are whole when young and become slightly split or broken at the edge when they reach adulthood. The flower buds develop from the buds of some of the leaves, which grow into long stems with a terminal inflorescence, which is the chapter (INFOAGRO, 2014).

The leaves also originate from the rhizome, arranged in a rosette shape, with a rounded outline in the early stages of development, which as they grow become slightly jagged at the edges, erect, elongated, about 40 cm long, lanceolate and supported by long petioles, from which the floral shoots often develop, which develop peduncles with a terminal inflorescence in a chapter. The ligulate flowers vary greatly in color, shape and thickness depending on the gerbera cultivar (MERCURIO, 2002; INFOAGRO, 2014).

The gerbera has a perennial cycle that can last for many years, however, it is recommended to grow it for a maximum of two to three years, as during this time the flowers produced still have commercial qualities (PEREIRA, 2013).

Currently, the preferred market demand in terms of petal color is as follows: red (22%), yellow (18%), lilac (17%), orange (14%), pink (13%) and other (16%) (INFOAGRO, 2014).

2.2 ECONOMIC IMPORTANCE

The commercialization of gerbera in North America began in 1920 (ROGERS; TJIA, 1990), with the aim of producing cut flowers. The wild species of *Gerbera jamesonii* and the cultivars developed initially had long stems and were better adapted for use as cut flowers than as pot flowers.

The production and consumption of flowers and ornamental plants in Brazil has been following the trend of expansion of the world market, which is also growing every year. Due to the high commercial value of its products and the fact that the production cycle for flowers and ornamental plants is relatively short, there is the possibility of a quick economic return, as well as the fact that the commercial value of floriculture products is usually high, compared to vegetables and fruit, for example (TERRA; ZÜGE, 2013).

The flower and ornamental plant segment is facing an increasingly demanding and competitive market, especially with regard to product quality standards, whose certification uses various criteria, including plant durability (ASGHARI et al., 2014). In addition, it requires technology, knowledge on the part of producers in all aspects of the plant, be it physiological, sanitary, in addition to the water or climate requirements for each species, these points being extremely relevant in the cultivation of flowers and ornamental plants.

In addition, Brazil has a very diverse climate, as most of the country has a tropical climate, but the south and southeast have a milder climate, making floriculture dependent on protected structures (protected environment) and appropriate cultivation techniques, especially for growing seasonal species. Despite these obstacles, the floriculture production chain has been lucrative, and in 2014 the sector's GDP was R$4.51 billion (IBRAFLOR, 2014).

According to data from the Brazilian Institute of Floriculture - IBRAFLOR (2014), Brazil has an area cultivated with ornamental plants of 15,000 hectares, with the state of Sao Paulo standing out as the main producer (7,000 ha), followed by Rio Grande do Sul (1,360 ha) and Santa Catarina (988 ha). There are 8248 producers in this segment, with the state of Sao Paulo standing out with 2288 producers, followed by Rio Grande do Sul and Rio de Janeiro with 1550 and 1030 producers, respectively (NEVES E PINTO, 2015). Menegaes et al. (2015) made a diagnosis of the retail flower market, concluding that gerbera is the second most commercialized cut flower among retailers in Santa Maria/RS.

This floriferous plant ranks fourth among the most consumed flowers in Brazil and is marketed mainly as a cut flower, with pot cultivation not being widespread (JUNQUEIRA; PEETZ, 2005; LUDWIG et al., 2010).

2.3 PRODUCTION AND QUALITY COMPONENTS

With regard to the marketing of gerbera, the criteria for its classification are: separating lots according to the length of the stem, which must be firm, without sharp deviations and hydrated, giving the flower support; based on length, the stems must have a uniform thickness of between 5 and 6 mm and finally the lots must be uniform in terms of the size of the bud (VEILING HOLAMBRA, 2016).

Gerberas can have defects that will define their commercial quality. The most uniform lots are classified as A1, while the A2 lot allows for a greater number of stems with defects, which can be severe (evolving during the post-harvest period) or mild (not evolving during the post-harvest period). Stems that show dehydration, botrytis, mildew, powdery mildew and mechanical damage should be avoided during selection, as these are serious defects that interfere with quality (VEILING HOLAMBRA, 2016). Flowers that are characterized by having a long stem, an inflorescence with an attractive structure and a variety of colours are attributes that are considered when selecting cut flowers (LUDWING et al., 2010).

As they meet these criteria, gerberas are mainly marketed as cut flowers, and their commercial demand has been widespread in recent years (SOLGI et al., 2009). However, like all cut flowers, they have a short post-harvest life due to the fact that senescence processes are triggered more quickly from the moment they are cut (RABIZA-SWIDER et al., 2016).

2.4 WEATHER AND WATER CONDITIONS FOR THE GERBERA CROP

Basic care is needed when managing the flower sector. As far as climate is concerned, gerberas are little influenced by photoperiod, but light helps to emit the lateral shoots that give rise to new chapters (LUDWIG et al., 2010). According to Lin and French (1985), sensitivity to photoperiod varies according to the variety.

The level of light, together with the temperature, has an influence on growth. Too much light and a high temperature imply accelerated growth to a certain extent (GUISELIN, 2002).

During the spring and summer periods, which are characterized by high light intensity and high temperatures, there is an increase in vegetative growth and a reduction in production quality, which is why it is important to use shading screens when growing gerbera (LUDWIG, 2007). The plants grow well in places without direct light, but require solar radiation of around 40,000 lux (GUERRERO et al., 2012).

When growing ornamental species, it is necessary to control the conditions of temperature, humidity, radiation, soil, wind and atmospheric composition, which is provided in protected environment conditions. The main characteristic of using a protected environment is that it avoids seasonality and allows the grower to have a diversity of products without interference from the weather.

Air temperature is a factor that influences the emission, leaf growth and earliness of flowering. Relative humidity between 75 and 90% is not a problem for the crop (INFOAGRO, 2012), but values above 70% during the day and 85% at night favor the occurrence of diseases such as botrytis and can cause spots and deformations on the flowers (PEREIRA, 2013).

2.5 GROWING IN A PROTECTED ENVIRONMENT

In Brazil, the predominant crops grown in protected environments are vegetables and ornamental plants, due to the greater sensitivity of these species to environmental conditions. Producers' desire to meet market demand has led to the expansion of protected growing in many regions of Brazil (BÄR, 2017), especially during unfavorable climatic periods.

In Portugal, greenhouse cultivation accounts for around 26% of the total cultivated area (BUAINAIN et al., 2007). Cultivation in a protected environment is an evolution in agricultural production, as it not only makes it possible to supply products throughout the year, especially seasonal crops, but also enables the geographical expansion of production centers to regions where climatic conditions are not favorable for the vast majority of crops.

Producers who grow crops in a protected environment do so because this way they obtain better quality products, thus increasing their productivity and making a product available to the market that would not be possible under field conditions (PEREIRA, 2013).

With regard to the cultivation of gerbera, the protected environment has become an alternative that has many advantages, including: guaranteeing the best quality of the product; reducing the occurrence of diseases that attack the aerial part; increasing the yield per area; facilitating the execution of cultural treatments; improving the production schedule, providing shorter cycles; reducing nutrient losses through leaching, resulting in the rational use of fertilizers.

The disadvantages include the high cost of setting up and maintaining the system and inadequate phytosanitary control, favoring the occurrence of pests and diseases, since many diseases in protected cultivation tend to become more severe when compared to conventional cultivation, because as well as the environment being favorable, the nutritional state of the plants, together with irrigation management and the greater density of plants, make the conditions more favorable to pathogens (VIDA et al., 1998; ZAMBOLIM et al., 1999; ZAMBOLIM et al., 2000).

2.6 SUBSTRATE CULTIVATION

Among the most popular crops grown in substrates are ornamental plants, which, due to their exoticism and beauty, are among the most valued on the national and international markets. However, little work has been done on the production of these species in substrates (ARAÚJO, 2010).

In Brazil, there is a wide variety of substrates with different physical and chemical characteristics, which can interfere with the development of gerbera (*Gerbera jamesonii* L.). Ludwig et al. (2010), in a study with gerbera, reported that the selection of an appropriate substrate for growing plants requires prior knowledge of its chemical and physical characteristics, which can interfere with plant growth.

The substrate must be better than soil in terms of characteristics such as water economy, aeration, permeability, buffering power for pH values and nutrient retention capacity. The use of water peat has advantages because it is a light material, free of toxic compounds, weed seeds, nematodes and pathogens. This substrate provides good aeration, high moisture and nutrient retention capacity. Peat has a high organic matter content, low pH value, high buffering power, high water retention capacity and good aeration, making it the main component for making substrates, and it is used as a benchmark for studying new materials (FERMINO, 2014).

Plants respond differently according to these characteristics. Knowledge of these factors contributes to the production of better quality plants that generate greater commercial value.

The physical and chemical characterization of substrates is necessary for their correct formulation and also for recommending and monitoring fertilizers in protected cultivation systems (ABREU et al., 2007). According to the results of studies on cut gerbera in the literature, the substrate must have a high water retention capacity, but at the same time it must have a large number of macropores to facilitate rapid drainage after irrigation (ROGERS; TJIA, 1990), with positive results for precocity, number of inflorescences per plant and uniform quality (MASCARINI, 1998).

Regarding the chemical characteristics, information was recorded on the pH of the substrate, which should be kept between 5.5 and 6.5 (ROGERS; TJIA, 1990), with a reduction in production when the pH is above 6.0 (SONNEVELD; VOOGT, 1997; SAVVAS; GIZAS, 2002).

According to Caballero et al. (2007), the availability of nutrients is one of the essential factors that influence the suitability of organic substrates for plant growth. This availability may not only depend on the composition of the substrate, but also on adsorption capacity, pH, biological stability and the presence of dissolved organic compounds.

Nowadays, a wide range of ornamental plants are grown in containers, which alter the relationship between the plant and the growing medium compared to open-air cultivation (FERMINO, 2014). In pots, the volume available to the roots is restricted and it is in this space that the plant's root system must satisfy its demands for air, water and nutrients. The growth medium used must therefore provide adequate storage of water and nutrients and provide good aeration (BEOZZI, 2013).

Rogers and Tjia (1990) pointed out that when growing gerbera, the oxygen potential in the soil is important and should generally be kept at a relatively high level. Ideally, the ratio of 1/3 substrate, 1/3 moisture and 1/3 air should be maintained both at the start and during cultivation.

The selection of the substrate should be based mainly on the physical characteristics for a given container and a given management and for the species to be grown in order to contribute to the best use of water and nutrients, determining the best quality of the final product (FERMINO, 2003).

2.7 IRRIGATION MANAGEMENT

Understanding the relationship between water and the water-soil-plant interaction system is essential, as it helps us to understand the mechanisms of interaction between the environment and the plant's response mechanisms to the environment (MENEGAES, et al., 2017). Irrigation is characterized by a practice that provides the efficiency of cultural growth and development, being necessary, especially in the development of crops in a protected environment, characterizing the only source of entry of the water resource (PADRÓN et al., 2015; GIRARDI, 2016).

In floriculture, according to Girardi (2012), irrigation determines the final quality of the product, especially in the ornamental sector. The sector is classified as sensitive to irrigation management, to excess or deficit, highlighting the importance of irrigation management (OLIVEIRA, 2016).

In order to improve cultivation in a protected environment, some authors have developed studies related to irrigation management, fertilization, humidity and temperature control, among other factors in this cultivation system. Noya et al. (2014) evaluated irrigation management in the cultivation of *Stenachaenium megapotamicum*, an ornamental plant which, like gerbera, belongs to the Asteraceae family, in a protected environment and concluded that the crop has the potential to be used in sustainable landscaping projects due to its ability to grow under conditions of water deficit.

Due to its importance as a resource, and with the preservation of its physical environment in mind, water must be optimized, especially in the agricultural sector (BERNARDO et al., 2006). It is essential to correlate water consumption, with a view to preserving it for major crops, where water demand is greater, and also in the ornamental sector, but for other purposes. The purposes of determining water consumption for the ornamental sector are related to the final product standard, the predominant characteristic of which is quality, obtained through precise irrigation management and specific cultural treatments for each crop.

Irrigation management consists of determining when, how much and how to apply water, within a broad concept, taking into account other aspects of the production system such as fertilization, phytosanitary control, climatological and economic aspects, crop management and strategies (TAVARES, 2007).

Authors have shown that irrigation increases the productivity of various vegetables (CARVALHO et al., 2004; DERMITAS; AYAS, 2009; VILAS-BOAS et al., 2008; ZENG; BIE; YUAN, 2009; BILIBIO et al, 2010), but the management of the irrigation system must provide suitable conditions to maximize the development and productivity of the crops. In addition, the rational management of an irrigation project must make it possible to increase the efficiency of water use and minimize investment and operating costs, so that the activity becomes economically viable and sustainable.

Irrigation can be managed in three ways: via the plant, the soil, the climate or a combination of these. It can be differentiated in the phenological stages of the crop according to its greater or lesser sensitivity to water stress and its effect on production (TAVARES, 2007).

Optimizing the irrigation schedule can lead to greater efficiency in the use of water, which is one of the most important objectives today. Adequate water and nutrient supply results in greater water and nutrient use efficiency, better production control and prevention of stress situations (RAVIV; BLOM, 2001).

In order to determine the quantity and availability of water for plants, it is necessary to know the dynamics of water in the soil. The water balance depends on evapotranspiration. This factor consists of presenting water conditions at the different stages of plant development (GOMES et al., 2008; SILVA, 2008), making it possible to show the amount of water consumed by the crop and determine the need for essential irrigation at the right time (LIBARDI, 2000; REICHARDT; TIM, 2004).

According to Ludwig et al. (2013) in a study with potted gerbera under fertigation rates and substrates, they used the gravimetric method to maintain soil humidity and observed that the percentage of water in the soil equivalent to the highest water availability (100% of the maximum water retention capacity in the soil) gave better quality plants.

According to Pereira et al. (1997) the water balance refers to soil water accounting, i.e. it consists of calculating the gain (inflow), loss (outflow) and storage of water in the soil over a period of time.

Water is supplied to the soil surface through precipitation and/or irrigation. The movement of water in the substrate is different from that in the soil, as it is generally a very porous material, its storage capacity is lower and its hydraulic conductivity is higher (GIRARDI, 2016).

2.7.1 Evapotranspiration

In protected environments, micrometeorological changes occur compared to the external environment and help to reduce water requirements and make the use of water by plants more efficient, due to the reduction in evapotranspiration (OLIVEIRA et al., 2014). In addition, evapotranspiration is lower in relation to the outside environment due to the diffusivity of the plastic cover and the conditions of temperature, relative humidity and the reduction in wind pressure, which are the main factors in the atmosphere's evaporative demand (DALMAGO et al., 2006; REIS et al., 2009; GUISELINI et al., 2010; PIVETTA et al., 2010; ANDRADE JÚNIOR et al., 2011).

According to Santiago et al. (2016), knowledge of the real water needs of crops is extremely important, since effective irrigation management is crucial in the management of water applied to agricultural production processes. Andrade et al. (2016) state that one of the ways to correctly manage irrigation is to estimate reference evapotranspiration (ETo) as a parameter.

The plant's water supply system does not always supply the amount of water needed to balance the flow of solar radiation energy with the energy spent on evapotranspiration (POLYSACK, 2007). Silva and Marouelli (1998) report that, in order to manage irrigation water correctly, it is essential to control soil humidity and/or know the evapotranspiration throughout the crop cycle. To do this, it is essential to know the parameters related to the plants, the soil and the climate so that the correct time and amount of water to be applied can be determined.

Evapotranspiration can be estimated by various methods, but the Penman-Monteith equation is a mathematical model recommended by the FAO as a standard method (ALLEN et al., 2006) for estimating reference evapotranspiration (ETo), as it incorporates aerodynamic and thermodynamic aspects, resistance to the flow of sensible heat and water vapor and resistance to water vapor transfer.

According to Bonachela et al. (2006) the methodology proposed by the FAO can also be applied to determine the evapotranspiration of crops grown in a protected environment.

2.7.2 Crop coefficient (Kc)

The crop coefficient (Kc) associated with the reference evapotranspiration (ETo) estimates the crop's evapotranspiration. Crop coefficients must be quantified according to the phenological stage of the crop, and can vary according to the location and climate in which they are being assessed (BARBOSA et al., 2015), which is why, according to Ribeiro (2006), this variation can alter the crop curve, and its main stages of development should be monitored. To solve this problem, meteorological variables have been used to relate Kc more directly to the crop's stage of development.

This strategy makes it possible to accurately determine the duration of the plant's phenological phases. In general, for most crops, Kc increases from a minimum value during germination to a maximum value, defined when the crop reaches full development, and tends to decrease from the start of maturity.

The value of Kc in a protected environment can be influenced by the interaction between the shape of the greenhouse structure, meteorological data and ETo estimation methods, so it is necessary to determine it for each crop, as well as for each growing location, making it an important tool in irrigation management and decision-making (RIBEIRO et al., 2009; DUARTE et al., 2010).

Therefore, knowledge of ETc and Kc during the crop's development stages is of fundamental importance for rational irrigation planning (ANDRADE et al., 2013).

A crop's water consumption is directly dependent on the atmospheric energy demand, the water content in the soil/substrate and the plant's resistance to losing water to the atmosphere. This makes it necessary to study supporting parameters for calculating the plant's water requirements, such as Kc, which is considered to be an indicator of physical and biological significance that is important in agricultural decision-making (OLIVEIRA et al., 2013).

Chapter 3

3. MATERIAL AND METHODS

This chapter includes the materials and methods used to carry out the experiment, the location, time and how the crop was managed, as well as the evaluations carried out during and after the experiment. Information was sought on the gerbera crop and its production requirements. All care and management was carried out in accordance with the recommendations.

3.1 PLACE AND TIME OF THE EXPERIMENT

This work was carried out in the municipality of Santa Maria, in the Floriculture Sector of the Polytechnic College, belonging to the Federal University of Santa Maria - UFSM, RS (29°43'23"S and 53°43'15"W, at an altitude of 95 m), from November 2016 to May 2017. The study region has a Cfa climate - humid subtropical, with hot summers and no defined dry season - according to the Koppen climate classification.

Temperatures throughout the year generally range from 10 °C to 31 °C and the average relative humidity is 74.4%.

The experiment was conducted in an air-conditioned greenhouse with a PadFan cooling system and hot air heating.

3.2 EXPERIMENTAL AREA - PLANT MATERIAL/VESSELS

The *gerbera* seedlings (*Gerbera jamesonii*) were purchased from the company Uniplant, located in Holambra - SP, which is the largest center for the production and distribution of flowers and plants in Latin America. The variety used was Caribá, which has red flowers (Figure 1). The seedlings were transplanted on November 3, 2016.

Rigid black plastic pots with a capacity of 25 liters were used, with a top opening diameter of 34 cm and a height of 38 cm, with drains at the bottom. One seedling was transplanted per pot, i.e. each pot was an experimental unit.

Figure 1 View of the cut gerbera flower, Caribá cultivar. Santa Maria - RS, 2017.

Source: (Piroli, 2016).

3.3 EXPERIMENTAL DESIGN

The experiment was conducted in a greenhouse measuring 600m^2 (20x30m), with a height of

3.5 meters. A sketch of the installation area can be seen below (Figure 2):

Figure 2 Sketch of the experimental area.

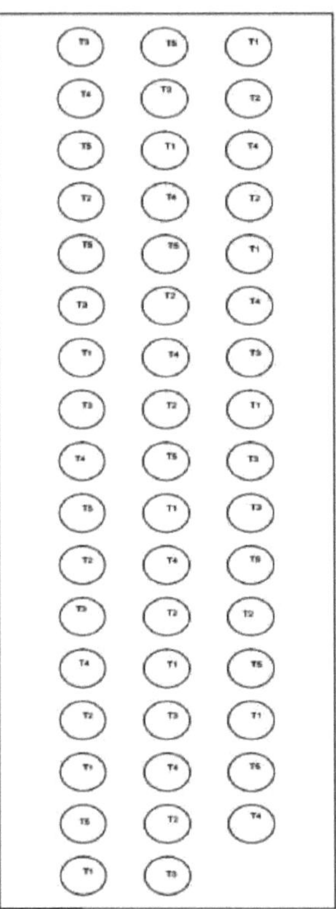

Source: (Piroli, 2016).

The experimental design was organized in a completely randomized design (DIC), with five treatments (water retention levels) and 10 replicates for each treatment, totaling 50 pots, with each pot representing an experimental unit (EU).

3.4 CHEMICAL AND PHYSICAL ANALYSIS OF THE SUBSTRATE

According to the chemical and physical analysis, we can check the attributes of the water peat substrate used in the experiment (Table 1).

The pH value was 6.50, which according to Fermino (2014) is within the ideal range for growing most plants, which is between 5.5 and 6.5 (pH in H_2O). The pH is an extremely important factor for the growth and development of plants, as it directly affects the availability of nutrients and can compromise the production and final quality of the product (LUDWIG et al., 2014).

Table 1- Chemical and physical analysis of the water peat substrate. Santa Maria - RS, 2017.

Indicators	Substrate values	Reference values
pH (H_2 0)	6,50	5,5 a 6,5

		0,36 -
EC (mS cm)$^{-1}$	0,78	0,65
DU (kg m)$^{-3}$	573,77	-
DS (kg m)$^{-3}$	257,36	350 a 500
UA (%)	55,15	-
PT (%)	88,46	80 a 90
AS (%)	16,03	10 a 15
AFD (%)	30,27	20 a 30
AD (%)	33,81	25 a 35
AT (%)	3,54	5
RA (%)	38,63	20 a 30
CRA 10 (%)	72,44	-

CRA 50 (%)	42,17	-
CRA 100 (%)	38,63	-

DU = wet density; DS = dry density; UA = actual humidity. pH = determined in water, diluted 1:5 (v/v); EC = electrical conductivity obtained in a 1:5 (v/v) solution PT = total porosity; EA = aeration space; AFD = easily available water; AT = buffering water; AD = available water; AR = Remaining water. CRA10 = water retention capacity under suction of 10 cm water column; CRA50 = water retention capacity under suction of 50 cm water column; CRA100 = water retention capacity under suction of 100 cm water column.
*average of three samples

The electrical conductivity value was 0.78 mS cm^{-1} , which is above the value considered ideal (0.36 - 0.65 mS cm^{-1}). High levels of salt in the soil or substrate can cause damage to plant growth and development.

Density represents the relationship between the mass and volume of the substrate and, in general, dry density is used as a determining parameter for evaluation, as wet density can vary depending on the amount of water in the sample at the time of analysis. Normally, the reference dry density values for substrates are 350 to 500 kg m^{-3} (CONOVER, 1967).

The ideal total porosity for substrates should be between 80 and 90% according to Verdonck and Gabriels (1988), so in the substrate evaluated, it can be seen that the content was within the ideal range. With regard to the values obtained for the aeration space (AS) of the substrates, the same authors suggest ideal values of between 10 and 15%, but the substrate showed a value above this range. High AS values can lead to water deficiency in plants, especially when irrigation is infrequent (ZORZETO et al., 2014).

For ideal values of available water, the contents are between 25 and 35 % (DE BOODT; VERDONCK, 1972; CATTIVELLO, 1991; SCHAFER et al., 2008), and the values found for peat are within the ideal range.

According to Schmitz (2002), the ideal readily available water content is between 20 and 30%; and for buffering water, which represents retained water, the ideal is around 5%. Although the remaining water is not available to most plants, its main relevance lies in its influence on the substrate's properties, such as electrical conductivity, thermal capacity and hydraulic conductivity, with 20 to 30% being considered ideal (SCHAFER et al., 2015).

The highest water retention capacity of the substrate was at the lowest water column pressure (10% of VC). This may show that the plant has more water available, making it easier to remove water from the substrate.

1.5 DETERMINATION OF VESSEL HOLDING CAPACITY (VC)

To determine the water retention capacity of the pot, the substrate was first dried and three 25-liter samples (pots) were taken, after which enough substrate was added to reach the edge of the pot. The plastic pots were filled with 7.8 kg of water peat (the mass of substrate needed to completely fill the pots), and the water retention capacity (WRC) was determined, according to the methodology described by Mello (2006) adapted by Schwab (2011), resulting in 14.500 kg. Once the pot's water retention capacity had been determined, treatments were obtained for values of 120%, 100%, 80%, 60% and 40%, which resulted in the different water availability limits:

$$PV_{120\%} = (PV_{crv} - PV_{seco}).1{,}2 + PV_{seco}$$

$$PV_{100\%} = (PV_{crv} - PV_{seco}).1 + PV_{seco}$$

$$PV_{80\%} = (PV_{crv} - PV_{seco}).0{,}8 + PV_{seco}$$

$$PV_{60\%} = (PV_{crv} - PV_{seco}).0{,}6 + PV_{seco}$$

$$PV_{40\%} = (PV_{crv} - PV_{seco}).0{,}4 + PV_{seco}$$

Where $PV\%$ is the weight of the pot for each of the treatments; $PVcrv$ is the water retention capacity; $PVseco$ is the weight of the pot filled with completely dry substrate.

The pots were randomly distributed and then the irrigation rates were applied according to each treatment. The weight of the pots for each treatment resulted in 15.825 kg for the 120% treatment; 14.500 kg for the 100% treatment; 13.175 kg for the 80% treatment; 11.850 kg for the 60% treatment and 10.525 kg for the 40% treatment of the maximum holding capacity of the pot.

1.6 WATER CONSUMPTION

The water consumption of the species was determined by reducing the humidity of the pot using weighing scales. Irrigation to supplement the weight of water to reach the maximum limits stipulated in each pot of each treatment was applied manually.

The crop's water consumption was determined using the water balance equation as shown in the following expression:

$$Etr = \sum_{i=1}^{L} Mi - \sum_{i=1}^{L} Mi + 1 + I - D$$

Where *Etr* is the actual evapotranspiration of the potted plant in a time interval *Át* of seven days; Mi is the mass of substrate and water contained in the pot at the beginning of the time interval (At) considered; *i* is the index representing the time interval (At) considered for the balance; $Mi+i$ is the mass of substrate and water remaining at the end of the time interval (At) considered; *I* is the irrigation applied to the pot in the time interval At and *D* is the percolation (or drainage) that may occur. The variation in water storage in the pot (M_i - M_i+i) was obtained by weighing the pots.

The variation in water storage in the pot (Mi - M_i+i) was determined by weighing the pots on a scale with a capacity of 50 kg. The established watering shift was seven days. The experiment was conducted in a greenhouse, so irrigation was the only way of supplying water to the plants, disregarding precipitation and surface runoff, as the pots were not saturated above their retention capacity.

Internal drainage or percolation only occurred in the pots with 100% and 120% VC. In this case, the amount of percolated water was measured and taken into account when calculating water consumption. The values obtained from the difference between the weighed mass established for each treatment (g day^{-1}) and the mass presented on the date of weighing minus percolation, were transformed into values corresponding to the water table (mm day^{-1}). Accumulated water consumption was the sum of all the water consumed during the period.

1.7 IRRIGATION

Irrigation was manual and based on the weights previously established for each pot at fixed intervals of seven days between irrigations, i.e. water was added weekly until the pre-established value for each treatment was reached. The experiment lasted six months. Phytosanitary control was carried out
according to indications for growing cut gerbera.

1.8 EVALUATIONS DURING THE EXPERIMENT

The crop coefficient for cut gerbera was determined, and the following variables were evaluated: stem length (cm); inflorescence diameter (cm); stem diameter (mm). These parameters are factors in the productivity and quality of the crop's flower stalks.

1.9 STATISTICAL ANALYSIS

The data was analyzed using SISVAR 5.6 software (FERREIRA, 2011), and the graphs were drawn up using the SigmaPlot 11.0 program. Analysis of variance was carried out at the 5% probability of error level and, as the data was quantitative, regression analysis was carried out and the maximum technical and economic efficiency for stem production was calculated.

1.10 CLIMATOLOGICAL DATA

A digital thermo-hygrometer was installed inside the greenhouse to check the maximum and minimum temperatures and relative humidity. Readings were taken every 15 minutes. Other data such as insolation and solar radiation were obtained from the automatic station of the National Meteorological Institute (INMET), installed at UFSM.

Reference evapotranspiration (Eto) was determined using the Penman method with the aid of Excel spreadsheets, which uses the following mathematical expression:

$$Eto = \frac{s}{\gamma} . \frac{Q^*}{2,45} + Ea$$

$$Ea = (0,2625 + 0,138.U_{2)}.(es - e)$$

$$s = \frac{41015\ es}{(T + 237,5)^2} = \frac{2505196}{(T + 237,5)^2}^{[7,5\frac{T}{237,5} + T]}$$

In which: *Eto:* is the reference evapotranspiration (mm day^{-1}); *s*: is the tangent of the saturation curve (hPa°C-1); *A:* is the psychrometric constant (0.66 hPa °C^{-1}); *Q**: is the radiation balance (MJ m^2 day^{-1} *; Ea:* is the saturation aerodynamic factor; *U: is the* wind speed at 2 meters (m/s-1); *es - e: is the* saturation deficit (d).

The crop coefficient (Kc) was determined for each crop cycle using the formula:

$$Kc = \frac{Etc}{Eto}$$

Where *Kc:* is the crop coefficient; *Etc:* is the crop evapotranspiration (mm day-1); *Eto* is the reference evapotranspiration (mm day-1).

1.11 DATA COLLECTION FOR THE WATER BALANCE

Determining the water consumption of the gerbera crop for the production cycle studied began on October 3, 2016, by weighing all the pots weekly (Thursday) (Figure 3). The reduction in the mass of the pots corresponded to evapotranspiration.

Figure 3 Weighing the pots on a scale. Santa Maria - RS, 2017.

Source: (Piroli, 2016).

Irrigations were carried out manually until each pot reached the mass limits stipulated for each treatment. The water lost was replaced evenly over the entire surface of the pot. The pots were placed on plastic plates to collect any drainage, which was measured and accounted for when calculating the water balance.

3.12 FERTIGATION

The main nutrients required by plants are: nitrogen (N), potassium (K), phosphorus (P), calcium (Ca), magnesium (Mg), sulphur (S), iron (Fe), manganese (Mn), zinc (Zn), copper (Cu), boron (B) and other nutrients.

The nutrients N, Ca, Fe and Mg are initially responsible for vegetative growth and the accumulation of reserves for future sprouting, while P and Ca are responsible for the sprouting of the buds that will form the stems. Therefore, larger amounts of nitrogen should be applied at the start of the vegetative phase, while the same should be done with potassium during the reproductive phase, as this will be responsible for the growth and coloring of the flowers and leaves.

Fertigation followed the recommendations for the gerbera crop, with soluble fertilizer with a formulation of N-P-K - 18-18-18 plus micronutrients according to the manufacturer (18% Nitrogen, of which 9.8% N-Nitric and 8,2% N-Amoniacal, 18% Phosphorus, 18% Potassium, 0.14% Iron, 0.05% Boron, 0.02% Copper, 0.08% Manganese, 0.008% Molybdenum, 0.05% Zinc) in the vegetative phase, when the gerbera needs more nitrogen and potassium.

At the beginning of the formation of the flowering stems, it was changed to the N-P-K - 6-12-36 formulation (6% Nitrogen, 4.5% N-Nitric and 1.5% N-Amoniacal, 12% Phosphorus, 36% Potassium, 1.8% Magnesium, 8% Sulphur, 0.07% Iron, 0.025% Boron, 0.01% Copper, 0.04% Manganese, 0.004% Molybdenum, 0.025% Zinc) until the end of the experiment, because according to Malavolta, Vitti and Oliveira (1997), potassium is important for stem length, stimulates vegetation and tillering in grasses, increases carbohydrate content, promotes the storage of sugars and starch, increases water utilization and resistance to drought, frost, pests and diseases, in addition, according to Medeiros et al., (2007), the relationship between the quantities of nutrients in the gerbera crop is fundamental for the quality and productivity of the flowers. The NPK ratio for gerbera in the initial phase is 2:0, 5:1, and 1:0.5:2 in the production phase, according to BioLab tecnologia Vegetal Ltda.

Fertilization was carried out in conjunction with the water replenishment of the pots, i.e. weekly (Thursday). During the cultivation cycle there was an attack by slugs and banana flies. To control this, we applied lesmicide (metarex) in the form of granules and evidence insecticide at a dosage of 1 g L^{-1} using a backpack sprayer (Figure 4).

Figure 4 Application of insecticide to control whitefly. Santa Maria - RS, 2017.

Source: (Piroli, 2017).

3.13 HARVESTING

The gerbera was harvested with pruning shears, one centimeter above the level of the substrate. Harvesting was continuous throughout the experiment, i.e. whenever new stems were at harvest point, they were harvested. The variables assessed in the gerbera plants occurred at harvest point, characterized by the opening of the stamens with the flowers showing at least two open circles with pollen release (LIN; FRENCH, 1985). The first flower stalks were harvested 60 days after transplanting the seedlings into their permanent pots. After harvesting, the following production components were assessed: stem length (measured from the point of insertion to the apex of the chapter, with the help of a graduated card and a ruler) (Figure 5); inflorescence diameter (from one end of the chapter to the other) measured with a graduated ruler (Figure 6).

Figure 5 - Evaluation of the stem length of the cut gerbera, Caribá cultivar, Santa Maria - RS, 2017.

Source: Piroli (2017).

Figure 6 - Evaluation of the chapter diameter of the cut gerbera, Caribá cultivar, Santa Maria - RS, 2017.

Source: Piroli (2017).

3.14 ECONOMIC PARAMETERS: MAXIMUM TECHNICAL AND ECONOMIC EFFICIENCY

An agricultural production response function is defined as one that expresses the physical relationship between the quantities used of a certain set of materials/inputs and the maximum physical quantities that can be obtained from the product, for a given technique used. The use of yield and net revenue response functions to analyze the results of agricultural analyses is widespread (FRIZZONE et al., 1995; BERTONHA et al., 1999; PEREIRA et al, 2003; FRIZZONE et al, 2005; MONTEIRO et al, 2006; MONTEIRO et al, 2007; PEIXOTO et al, 2007; MARQUES SILVA et al, 2008; CARVALHO et al, 2009; DELGADO et al, 2012). Generally speaking, the problem is to find an optimal solution for the output/input combination that maximizes output and net revenue separately, subject to pre-fixed resource constraints and a given cost and price structure.

Efficient irrigation management requires information on the crop's water needs and the crop's water production function. In the specific case of gerbera, production is based on the number of stems produced over a six-month period (experimental period). The use of response functions makes it possible to find useful solutions for optimizing irrigation and obtaining maximum production.

Research involving factors such as water lamination and product, point to recommendations aimed at maximum physical productivity, not considering the economic aspect, a factor that should be taken into account, as the economic optimum does not always correspond to maximum physical productivity.

To obtain the yield function, regression analysis was used between stem yield and water applied, adjusted by a second-order polynomial model (Hexem & Heady, 1978):

$$Y = a + bX + cX^2$$

Where: *Y:* number of stems; *X*: water applied (mm); *a, b and c*: equation parameters.

With the adjusted stem production function, the blade corresponding to the maximum technical efficiency (X_{MET}) was determined:

$$X_{MET} = -\frac{b}{2C}$$

Production profit was obtained from the difference between the total monetary value of stem production and water costs:

$$L = Y.Py - X.Px$$

Where: L: lucro (R$); Y: number of stems, *Py*: product price (stems), *X*: applied water (mm); *Px*: water factor weight (R$ mm).$^{-1}$

The maximum economic efficiency (X_{MEE}) was obtained by calculating the first-order derivative of the equation for the number of stems, obtaining the equation that corresponds to the optimum blade that maximized revenue:

$$X_{MEE} = \frac{\frac{Px}{Py} - B}{2C}$$

Where: *Px:* water price and *Py*: product price (stems).

Chapter 4

4. RESULTS AND DISCUSSION

This chapter deals with the results obtained in this experiment, through data analysis, which were discussed and compared with other studies already carried out in the literature.

4.1 WATER CONSUMPTION

During the experiment, the average air temperature was 28 °C and the average minimum and maximum temperatures were 19.5 °C and 36.5 °C, respectively (Figure 7). Very high temperatures (greater than 29 °C) during the initial flowering period result in more leaves below the flower, producing taller plants with more internodes (GRUSZYNSKI, 2001).

The appropriate temperature for gerbera is 28 °C during the day and 20 °C at night (INFOAGRO, 2012). It can therefore be seen that the average minimum and maximum temperatures are outside the ideal range and the average temperature value obtained was within the ideal range for cultivation.

The average relative humidity was 68.3% and the minimum and maximum humidities were 61.9 and 74.3%, respectively. Relative humidity should be between 75 and 90% so as not to pose a problem for the crop (INFOAGRO, 2012). Therefore, the maximum relative humidity found is conducive to the development of the gerbera crop. However, the same cannot be said for the minimum relative humidity or the average relative humidity.

Figure 7 Climatic data and water consumption of the gerbera crop at the different irrigation rates. Santa Maria - RS, 2017.

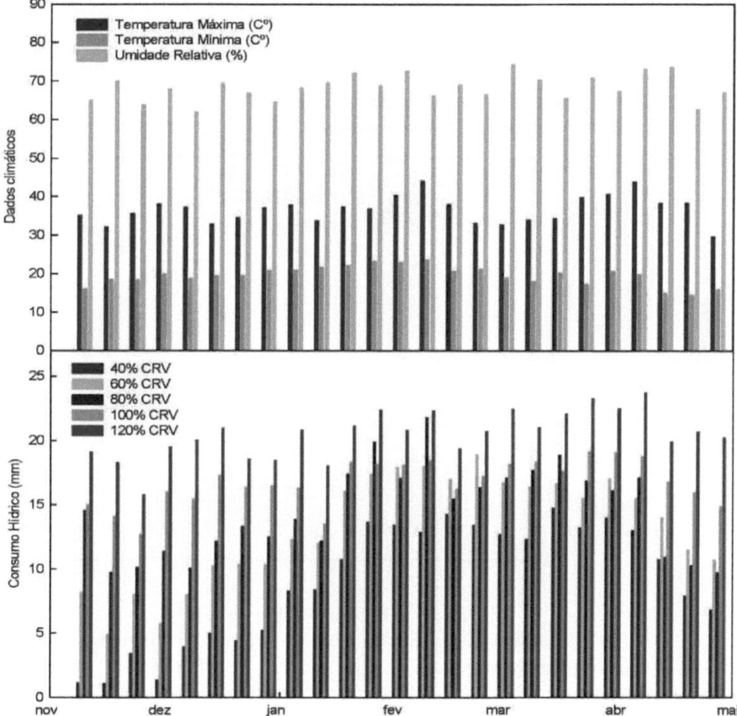

For the water consumption results for the gerbera crop, considering a significance level of 5%, there was a statistical difference at 5% in relation to water availability. This variable fitted the quadratic regression model, in which the highest volume of water consumed throughout the crop cycle was obtained at 120% of the maximum water retention capacity.

Schwab et al. (2013) observed in their study of carnation crops subjected to four irrigation strategies (100%, 80%, 60% and 40% of water retention capacity) and two pot sizes (8 and 18 liters) that for the 18-liter pots, water consumption at 100% of water retention capacity was much higher than for the other treatments. These results can be explained by the greater availability of water, which led to an increase in exchanges between the plant and the atmosphere, characterizing the expansion of plant tissues and the development of plants in the absence of water factors that could cause the stomata to close and, consequently, a reduction in water consumption. These results can be seen in the treatments with the lowest irrigation rates, in which the lack of water caused the plants to stop growing and developing and the stomata to close, resulting in a reduction in water consumption and production (FERRAZ et al., 2012).

Higher water consumption does not necessarily imply greater plant development or productivity, since the highest green and dry mass values were obtained when the water content was less than 80% of the pot's maximum retention capacity. According to Farias et al. (2005) the highest water consumption of chrysanthemums was observed at the lowest soil water tensions, however, the plants grown under these tensions did not show the greatest development. This behavior can be explained by the relative growth rate, in which plants in less favorable soil water conditions have greater adaptive capacity compared to those that grew in more satisfactory water availability.

Therefore, when studying irrigation, it is important to evaluate water consumption in conjunction with other factors, such as soil/substrate conditions, plant growth and development and climatic conditions, adapting to each particular situation. In this way, it will be possible to carry out more efficient management, applying the ideal amount of water to the plant.

In relation to the accumulated evapotranspiration of the gerbera crop for the different water availability, it was found that the treatment with 120% of the vessel retention capacity had the highest water consumption, as it was the treatment with the highest crop evapotranspiration, resulting in greater water replenishment than the other treatments (Figure 8).

Figure 8 Accumulated evapotranspiration of the gerbera crop for the different irrigation rates. Santa Maria - RS, 2017.

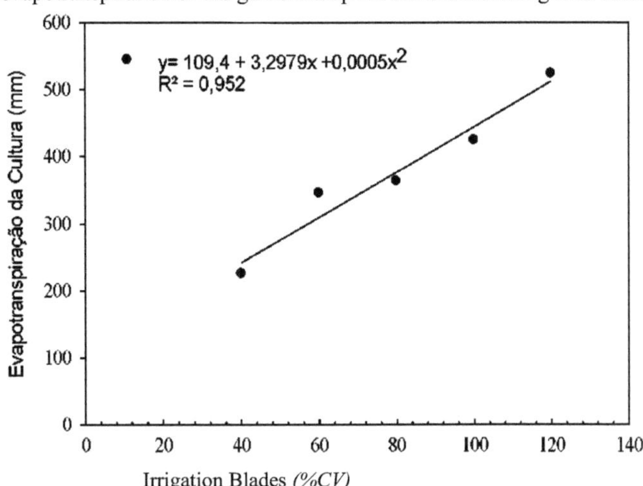

According to Girardi et al. (2016) in a study carried out on alstroemeria in a protected environment, subjected to different water availability (30, 45, 60, 75 and 90%) of the pot's holding capacity; when the pot's water conditions are kept at maximum holding capacity, the water moves more easily, preventing the plants from transpiring or evaporating the water contained in the substrate, resulting in higher water consumption. Thus, in an unsaturated substrate, the hydraulic conductivity is lower because only a fraction of the substrate is occupied by water, corroborating the data observed in this study.

8.2 CROP COEFFICIENT (KC)

Cut gerbera does not have a detailed phenological description with the distribution of its different phases during the production cycle in a protected environment using a peat substrate. To this end, a phenology was created, starting with the treatment of 100% water availability, with observations of the behavior of the plants identified within the interval after the gerbera seedlings were transplanted during the experiment. A proposed division of plant development with the respective average values of the crop coefficient (Kc) according to the phenological phases is presented below (Table 2).

Table 2- Days after transplanting (DAT), development stages, crop coefficient (Kc), reference evapotranspiration (Eto) for the cut gerbera cultivar Caribá. Santa Maria - RS, 2017.

Days after transplant (DAT)	Phases	Kc	Eto
0-29	Vegetative I	0,72	2,91
29-58	Vegetative II	0,81	2,92
58-87	Flowering I	0,85	3,18
87-116	Full bloom II	0,89	3,22
116-145	Full bloom III	1,33	2,58
145-174	Full bloom IV	1,06	1,81

An average of the Kc values for the different phenological phases of the crop gives an average value of 0.76 during the vegetative phase and 1.03 during the productive phase. There is little research information on determining the Kc of ornamental plants. Gomes et al. (2006), in a protected environment for heliconia, obtained average values of 0.88 and 1.23 for the vegetative and flowering (productive) phases respectively. Similarly, Gomes et al. (2008) obtained average values of 0.72 and 1.07 for the aforementioned phases of the alpinia crop.

Average crop coefficient values equal to 0.75 and 1.18 for the rose crop (cv. Carola), in the vegetative and productive development phases, respectively, were found by Oliveira et al. (2014) in protected environment conditions in Minas Gerais.

Girardi et al. (2016) in a study with the alstroemeria crop created a phenology, from the treatment of 90% of the pot's holding capacity, also in a protected environment in Santa Maria - RS,

found average values of 0.39,0. 41,0. 95, 1.50 and 0.75, in the vegetative phase, beginning of the
flowering, flowering, full flowering and fall flowering respectively.

The crop coefficient values obtained in this study may be correlated with the different water requirements during the phenological stages of gerbera in the greenhouse, which have not yet been determined by other studies. In this way, there is a need for further research into the development stages of the crop in question, taking into account the conditions under which it grows.
and pointing out that Kc values also change with the cultivar, irrigation system, ET estimation method adopted and crop management.

8.3 PRODUCTION PARAMETERS

8.3.1 Average stem length and stem diameter

The results show that the average length of cut gerbera stems was greater under conditions of 80 and 100% water availability (Figure 9). However, the flower stems of all the treatments meet the marketing standard for cut gerbera, where the length of the stems (CH) of gerberas adopted for marketing is CH>30 cm for classification in category A2 and CH>45 cm for category A1, according to IBRAFLOR (2017). Thus, despite the fact that the 40, 60 and 120% treatments did not differ
statistically, all the flower stalks are classified in the A2 category, i.e. all the water available showed satisfactory results for the average length of the stalks, which is a very important standard as it determines the uniformity of the batch.

Figure 9 Average length of stems (cm) in the different laminas (%CV). Santa Maria - RS, 2017.

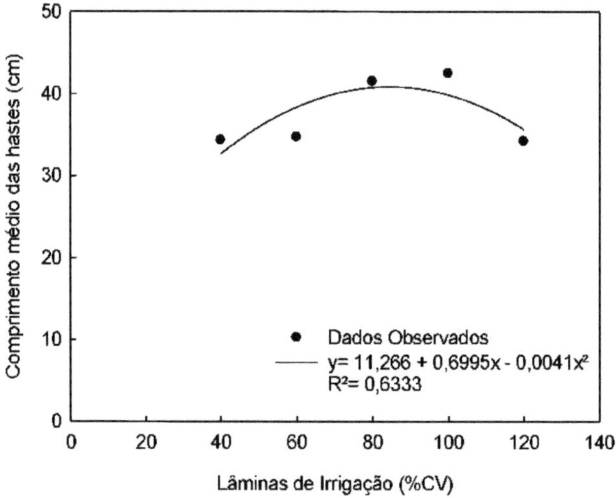

According to Silva (2017), evaluating the water demand and production of gerbera hybrids as a function of different water laminas, in the Submedium of the São Francisco Valley, in a greenhouse, they observed that excess and deficit water had a negative influence on the stem length of the DTCS hybrid, This was not confirmed for the Essandre hybrid, which showed a decrease only when subjected to a 40% deficit, satisfying the data observed for the length of the floral stem, which was negatively affected by water deficit and excess water throughout the experiment period.

Similar results were observed by Pereira et al. (2003), when studying levels of water replenishment in potted chrysanthemum cultivation belonging to the same family as gerbera, in a protected environment, they found that the greatest length of floral stems was found in treatments in which the crop did not suffer water deficit, similar responses were found by Girardi et al. (2017), who observed greater height of the stems with the increase in water availability in the alstroemeria crop, these results are in line with those obtained in the present study.

There needs to be a balanced relationship between stem diameter and plant height for both pot-grown and bed-grown gerberas, as taller and thinner stems are prone to breakage due to poor support, influencing the aesthetic quality and interfering with the marketing process of the stems (LUDWIG et al., 2010).

Cut gerbera stems must be uniform in thickness, between 5 and 6 mm. This classification is important because it establishes the support of the stem and determines the uniformity of the batch, according to IBRAFLOR (2017). The 80% water availability was the one that gave the best results in terms of stem diameter for cut gerbera (Figure 10). However, all the treatments tested showed positive results, with standard diameters for commercialization.

Tsirogiannis et al. (2010), evaluating the effect of irrigation management in a greenhouse with 4-liter pots, observed that stem length and stem diameter are very important factors for the gerbera crop (cv Balance), as they define the crop's introduction into the cut flower market.

The study by Pereira (2013) and Fanela et al. (2006), evaluating the levels of water tension in the soil in gerbera cultivation, observed that the size of the flower stem and the number of chapters showed satisfactory responses in the treatment in which the plants did not suffer water stress.

Figure 10 Average diameter of the stems (mm) in the different laminas (%CV). Santa Maria - RS, 2017.

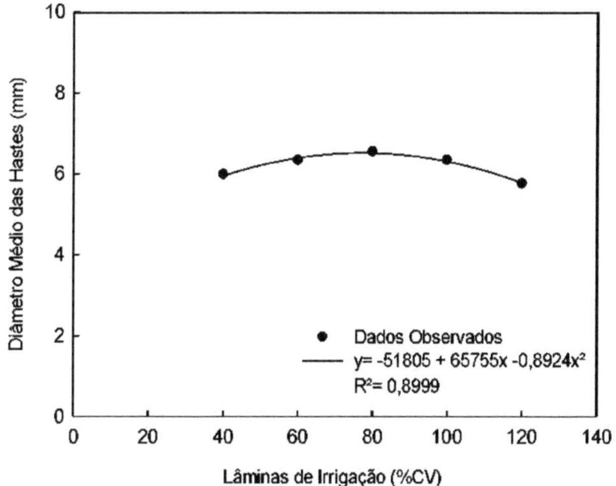

According to Neto (2017), in a study with the lisianthus crop, testing moisture limits and potassium concentrations in the soil solution when growing in a protected environment, he observed a reduction in the average diameter of the stems as a result of the reduction in soil moisture.

Stems with larger diameters are more rigid, which reduces tipping and breakage during harvest and post-harvest. It also allows for a longer shelf life due to the greater number of xylem vessels which allow for a greater flow of water, delaying the degeneration caused by water stress and allowing for greater post-harvest longevity (DOORN; WITTE 1991), which is an important characteristic when it comes to cut flowers.

4.3.2 Inflorescence diameter

The average diameter of the complete inflorescence was significantly influenced by the different pot retention capacities. The largest diameters were found in the treatments of 100, 80 and 60%
respectively, while the 40 and 120% treatments had smaller diameters (Figure 11). However, all the diameters found in the different water availability treatments were satisfactory for marketing (8 to 10 mm) and could be marketed according to the IBRAFLOR classification criteria.

Both a lack of and excess irrigation can have a negative effect, making nutrients unavailable to the plants, as excess water can cause nutrients to leach from the substrate, depleting it and making it unavailable to the plants, and can also interfere with oxygen restriction for the roots. A lack of water can affect the productivity of the plants and the characteristics of the flowers.

Tahir et al. (2002), after testing 25 sunflower genotypes subjected to water stress in Iran, observed a reduction in plant height (6.42%), leaf area (25.56%), chapter diameter (15.21%), weight of 1000 achenes (22.63%), yield per plant (34.13%) and stem dry weight (19.56%) in the treatments with the lowest available water content in the soil.

Figure 11 Average diameter of the inflorescences in the different laminas (%CV). Santa Maria - RS, 2017.

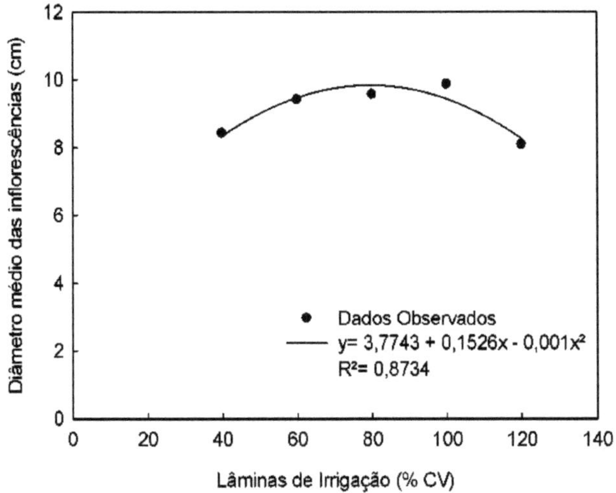

Dourado (2015), growing sunflower in a greenhouse in Mato Grosso, evaluated five doses of K2O (0, 40, 80, 160 and 240 mg dm^{-3}) and four levels of water replenishment in the soil (75, 100, 125 and 150% of field capacity), found that the largest diameter of 23.29 cm was observed at the highest level of water applied (150% of field capacity), with an increase of 28% when compared to the lowest replenishment rate (75% of field capacity). This corroborates the results found in this study, in which the diameter of the inflorescence was negatively influenced by the deficit and excess of available water.

4.3.3 Number of stems

It can be seen that the highest number of floral stems (NHF) was achieved when subjected to 80% water availability, producing an average of 11 stems/plant during the experiment period, while the other treatments provided reduced amounts of stems (Figure 12).

Similar results were found by Girardi et al. (2017), who evaluated the productivity and effects of different irrigation rates on the productivity of the alstroemeria crop in a protected environment, and observed that the highest number of stems was found at an irrigation rate of 90% of the pot's retention capacity.

Figure 12 Number of stems in the different blades (CV%). Santa Maria - RS, 2017.

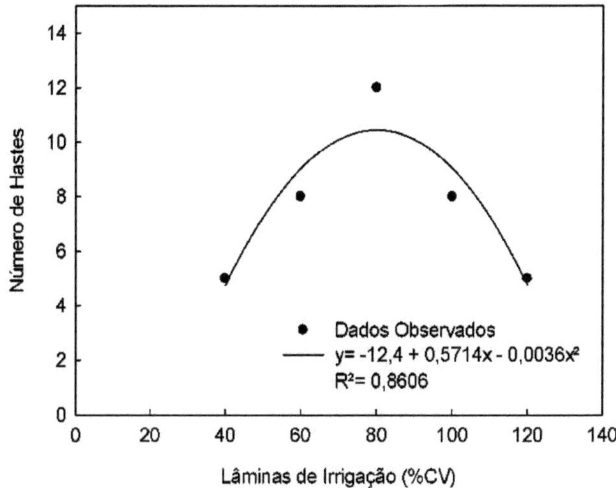

Pereira et al. (2016), evaluating the development of gladiolus produced in different concentrations of wood ash and humidity, found that the highest production of inflorescences was obtained under conditions with greater water availability in the soil. These results highlight the importance of studies on adequate water availability in the production of cut flowers, thus corroborating this research.

Similar results were found by Borges (2005), who, in a study with gladiolus evaluating the yield, quality and precocity of irrigated gladiolus in Bahia, observed a greater number of extra stems in the gladiolus crop with the highest irrigation rate applied. The same result was verified by Casarini (2000), evaluating the effects of different irrigation rates on the productivity of the rose crop, in a protected environment in Ceará, where he observed a better quality of roses as the irrigation rate increased.

4.4 MAXIMUM TECHNICAL AND ECONOMIC EFFICIENCY

The point of maximum technical efficiency (MET) was observed at 79.36% of pot capacity and was determined using the stem yield function.

Maximum economic efficiency (MEE) was obtained from the equation for the optimum blade that maximized revenue. Thus, the blade that corresponded to the MEE was 70.6 mm.

Among the mathematical models that describe a production function, the most commonly used for economic analysis in agricultural research are: Quadratic, Square Root, Mitscherlich and Power 3/2 (HEXEM & HEADY, 1978). However, the quadratic polynomial model, used by several researchers (TARSITANO; HOFFMANN, 1985; FRIZZONE, 1987; OLIVEIRA, 1993; PEREIRA, 2005), is most often the one that best represents the production estimate, allowing an analysis that defines maximum economic efficiency, using maximum productivity or maximum profit.

The value of stems varies according to the time of year (commemorative dates) and the way in which they are sold. The profit for different prices of stems (1.00, 2.00 and 2.50 R$), combined with the prices between the water factor and the product (Px/Py) (Table 3) is extremely important and can act as an indicator for assessing situations of economic viability or unviability.

Economic viability in the production process of species of economic interest in landscaping, such as cut gerbera, is one of the most efficient ways of avoiding losses. The price values for stems were based on the wholesale and retail prices surveyed in the region under study.

The wholesale flower trade sells its products to legal entities, i.e. those with a CNPJ, such as institutions, companies, restaurants and other shopkeepers who work in retail. This way, wholesalers offer products in large quantities (dozens or bundles) and at lower prices, as they sell the goods directly from the producers.

In retail, such as flower shops, prices are higher, but the end consumer has the option of paying in installments by credit card, for example.

Table 3- Revenue, expenses and profit for different types of stems according to the maximum technical efficiency and number of stems produced. Santa Maria - RS, 2017.

PX/PY	X MEE		NH	Sales prices for rods								
				1,00	2,00	2,50	1,00	2,00	2,50	1,00	2,00	2,50
%	**%**	**Mm**		**Revenue (R$)**			**Expenses (R$)**			**Profit (R$)**		
0	79,4	70,6	10,3	10,27	20,5	25,7	0,0	0,0	0,0	10,3	20,5	25,7
0,01	78	69,4	10,3	10,27	20,5	25,7	0,7	1,4	1,7	9,6	19,1	23,9
0,02	76,6	68,2	10,2	10,25	20,5	25,6	1,4	2,7	3,4	8,9	17,8	22,2
0,03	75,2	66,9	10,2	10,21	20,4	25,5	2,0	4,0	5,0	8,2	16,4	20,5
0,04	73,8	65,7	10,2	10,16	20,3	25,4	2,6	5,3	6,6	7,5	15,1	18,8
0,05	72,4	64,5	10,1	10,10	20,2	25,2	3,2	6,4	8,1	6,9	13,8	17,2
0,06	71	63,2	10,0	10,02	20,0	25,1	3,8	7,6	9,5	6,2	12,5	15,6
0,07	69,6	62	9,9	9,93	19,9	24,8	4,3	8,7	10,8	5,6	11,2	14,0
0,08	68,3	60,7	9,8	9,83	19,7	24,6	4,9	9,7	12,1	5,0	9,9	12,4
0,09	66,9	59,5	9,7	9,71	19,4	24,3	5,4	10,7	13,4	4,4	8,7	10,9
0,1	65,5	58,3	9,6	9,58	19,2	23,9	5,8	11,7	14,6	3,8	7,5	9,4
0,11	64,1	57	9,4	9,43	18,9	23,6	6,3	12,5	15,7	3,2	6,3	7,9
0,12	62,7	55,8	9,3	9,27	18,5	23,2	6,7	13,4	16,7	2,6	5,2	6,4
0,13	61,3	54,6	9,1	9,10	18,2	22,7	7,1	14,2	17,7	2,0	4,0	5,0
0,14	59,9	53,3	8,9	8,91	17,8	22,3	7,5	14,9	18,7	1,4	2,9	3,6
0,15	58,5	52,1	8,7	7,60	17,4	21,8	7,8	15,6	19,5	-0,2	1,8	2,2
0,16	57,1	50,9	8,5	7,35	17,0	21,2	8,1	16,3	20,3	-0,8	0,7	0,9
0,17	55,8	49,6	8,3	7,09	16,5	20,7	8,4	16,9	21,1	-1,3	-0,3	-0,4
0,18	54,4	48,4	8,0	6,82	16,0	20,1	8,7	17,4	21,8	-1,9	-1,4	-1,7
PX/PY* ratio between water factor and product NH* number of stems												

X MEE* blade corresponding to maximum technical efficiency (%) and economic efficiency (mm)

According to the table, it can be seen that the lower the ratio between stem length and irrigation rate, the greater the profit achieved. According to Oliveira et al. (2016), in a technical and economic evaluation of the effect of different irrigation rates and doses of nitrogen on the productivity of rose bushes grown in a protected environment, they found that the optimum rate tended to decrease as the ratio between the price of water and the price of a dozen roses increased. They consider this to be a trend, because by varying the price of water and keeping the price of roses fixed, the total economic rate of irrigation to be applied should be lower, so that the producer can obtain maximum profit from the activity.

Profit increases for all values of stems used (1.00, 2.00 and 2.50 R$) as the ratio between the water factor and the product decreases (Figure 13).

Figure 13 Profit generated, considering different stem lengths (Px) and recommended blades (Py). Santa Maria - RS, 2018.

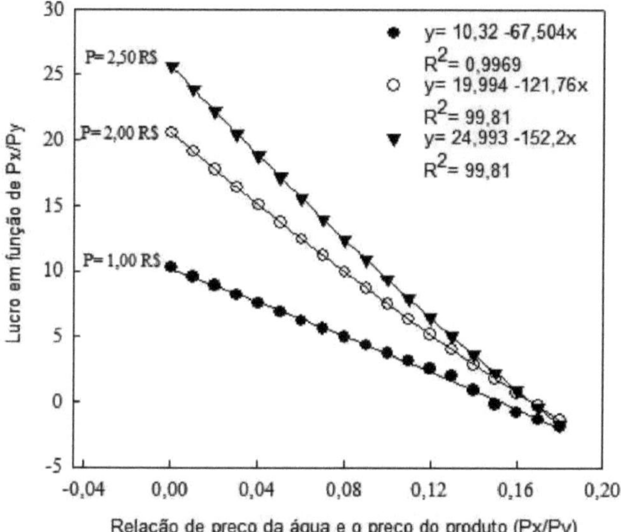

It can be seen that the final profit for stems of 1.00, 2.00 and 2.50 reais is initially positive and decreases as the Px/Py ratio increases. From the point at which the profit showed negative values, indicating that it is not economical to recommend the use of this amount of water.

The flower sector in Brazil has countless marketing advantages, in addition to the ease of adding value, the ability to differentiate the final product and the specificity of the product, which translates into a huge competitive advantage over other sectors. There are currently three sales lines that can be explored: value-added (flower shops) and mass consumption, which includes garden centers, supermarkets and the informal market (producers).

CONCLUSIONS

With regard to water availability, the development and production of gerbera plants showed the best results when grown in the range of 80% to 100% of maximum pot retention capacity.

The crop coefficient of gerbera grown in a protected environment varied from 0.72 in vegetative period I, 0.81 in vegetative period II, 0.85 for flowering period I, 0.89 for full flowering period II, 1.33 for full flowering period III and 1.06 for full flowering period IV.

The maximum technical efficiency for the production of gerbera stems was observed with a water table of 79.3% of the pot's holding capacity.

The maximum economic efficiency was found for the 70.6 mm blade.

Profit increases for all values of stems used (1.00, 2.00 and 2.50 R$) as the ratio of water to product decreases.

BIBLIOGRAPHICAL REFERENCES

ABREU, M.F.; ABREU, C.A.; SARZI, I.; PADUA JUNIOR, A.L. Aqueous extractors for the chemical characterization of plant substrates. **Horticultura Brasileira**, Brasília, v. 25, n. 2, p. 184-187. 2007.

ALLEN, R. G.; PEREIRA, L. S.; RAES, D.; SMITH, M. **Crop evapotranspiration**: Guidelines for determining crop water requirements. Rome: FAO, 2006. 298p. FAO Water and Drainage Study, 56.

ANDRADE JÚNIOR, A. S.; DAMASCENO, L. M. O.; DIAS, N. da S.; GHEYI, H. R.; GUISELINI, C. Climate variations in greenhouse cultivated with gerbera and relationship with external conditions. **Engenharia Agrícola**, v.31, p.857-867, 2011.

ANDRADE, A. D.; MIRANDA, W. L., CARVALHO, L. G. de.; FIGUEIREDO, P. H. F.; SILVA, T. B. S. da. Performance of tank coefficient calculation methods for estimating reference evapotranspiration. **Revista Irriga**, Botucatu, v. 21, n. 1, p. 119-130, 2016.

ANDRADE, A. R. S. de; CRUZ, A. F. da S.; CAVALCANTE, E. C. dos S.; ALBUQUERQUE, J. C. F. de; SOUZA, W. M. de. Estimation of evapotranspiration and crop coefficients for different stages of watermelon development. **Revista Brasileira de Geografia Física**, v.6, n.5, p.1417-1429. 2013.

ARAÚJO E.M; OLIVEIRA J.B; ARAÚJO E. M; LEDO E. R. F; MAIRTON G. DA SILVA M. G. Performance of eto estimation methods correlated with the standard Penman Monteith fao56 equation, in cities in the state of ceará. **Revista ACTA Tecnológica - Revista Científica** - ISSN 1982-422X, Vol. 5, número 2, jul-dez. 2010.

ASGHARI, R.; SALARI, A.; GHAREHDAGHI, S. Effect of pulsing solution and packaging type under exogenous ethylene on physiological characteristics and post harvesting quality of cut roses (Rosa hybrida). **American-Eurasian Journal of Agricultural & Environmental Sciences,** v.14, n.4, p. 329-335, 2014.

BEOZZI, S. **Valorization of organic waste in the formulation of alternative substrates to peat for the production of potted aromatic plants in organic production**. 2013. 74 p. Dissertation (Master's Degree in Environmental Engineering) - Technical University of Lisbon, Lisbon, 2013.

BÄR C. S. L. L. **Water availability and plant ash in the cultivation of potted gerberas**. 2017. 63p. Dissertation (Master's in Engineering Agricultural). Federal University of Mato Grosso. Mato Grosso. 2017.

BARBOSA, B. D. S.; OLIVEIRA, F. G.; DE FIGUEIREDO, E. P. Determining the crop coefficient (kc) of irrigated Tanzania grass in northern Minas Gerais. **Revista Irriga**, Botucatu, p. 11-20, 2015. Irriga & Inovagri Special Issue.

BERNARDO, S.; SOARES, A. A.; MANTOVANI, E. C. **Irrigation Manual**. 8ed. Viçosa: Ed. UFV, 2006.

BERTONHA, A.; FRIZZONE, J. A.; MARTINS, E. N.; Irrigation and nitrogen fertilization in pear orange production. **Acta Scientiarun**, Maringá, v.1, n.3, p. 537-542, 1999.

BILIBIO, C. et al. Vegetative and productive development of eggplant subjected to different soil water tensions. **Revista Brasileira de Engenharia Agrícola e Ambiental,** Campina Grande, v. 14, n. 07, p. 730735, 2010.

BONACHELA, S.; GONZÁLEZ, A. M.; FERNÁNDEZ, M. D. Irrigation scheduling of plastic greenhouse vegetable crops based on historical weather data. **Irrigation Science**, v.25, p.53-62, 2006.

BORGES, E. M. **Yield, quality and precocity of irrigated gladiolus in the Recóncavo baiano**. 2005. 64p. Master's dissertation - Federal University of Bahia. Bahia. 2005.

BUAINAIN, A. M.; FILHO, H. M. de. S.; GUANZIROLI, C.; BATALHA, M. O. **Family farming and technological innovation in Brazil: characteristics, challenges**

and obstacles. Campinas: Unicamp, 2007. p. 129-159.
CABALLERO R; ORDOVÁS J; PAJUELO P; CARMONA E; DELGADO A. Iron chlorosis in gerbera as related to properties of various types of compost used as growing media. **Communications in Soil Science and Plant Analysis** 38: 2357-2369. 2007.
CARVALHO, D. F.; DELGADO, A. R. S.; OLIVEIRA, R. F.; SILVA, W. A.; FORTE, V. L.; Maximizing production and agricultural revenue with water and nitrogen limitations using the interior point method. **Engenharia Agrícola**, Jaboticabal, v.29, n.2, p.321-327, 2009.
CARVALHO, J. de. A.; SANTANA, M. J. de. PEREIRA, G. M.; PEREIRA, J. R. D.; QUEIROZ, T. M. de.. Water deficit levels at different phenological stages of the eggplant crop (*Solanum melongena* L.). **EngenhariaAgrícola**, Brasília, v. 24, n. 2, p. 320-327, 2004.
CASARINI, E. **Irrigation management for rose cultivation in a protected environment**. Piracicaba 2000. 66p. Master's dissertation-
Luiz de Queiroz College of Agriculture.
CATTIVELLO, C. Physical properties in commercial substrates and their relationships. **Acta Horticulturae**, Wageningen, v.294, p.207-214, 1991.
CONOVER, C.A. Soil amendments for pot and field grown flowers. **Florida Flower Grower**, Florida, v.4, n.4, p.1-4, 1967.
DALMAGO, G. A.; HELDWEIN, A. B.; NIED, A. H.; GRIMM, E. L.; PIVETTA, C. R. Maximum evapotranspiration of the pepper crop in a plastic greenhouse as a function of solar radiation, temperature, relative humidity and air saturation deficit. **Ciência Rural**, v.36, p.785-792, 2006.
DE BOODT, M.; VERDONCK, O. The physical properties of the substrates in horticulture. **Acta Horticulturae**, Wageningen, v.26, p.37-44, 1972.
DELGADO, A. R. S.; DA SILVA ARAÚJO, W; CARVALHO D. F.; DO FORTE LEAL, V. P.; Planning irrigated agriculture in the North of Rio de Janeiro, using different mathematical programming techniques. **Revista Operational Research for Development**, v. 4, n.3, p.249-256, 2012.
DERMITAS, C.; AYAS, S. Deficit irrigation effects on pepper (*Capsicum annuum* L. Demre) yield in unheated greenhouse condition. **Journal of Food,Agricultural and Environment,** Helsinki, v. 7, n. 3/4, p. 989-1003, 2009.
DOORN, W.G.; WITTE, Y. Effect of dry storage on bacterial counts is stems of cut Rose flowers. **HortScience**, Alexandria, v.26, n.2, p.1521-1522, 1991.
DOURADO, L. G. A. **Irrigation and potassium fertilization in sunflower growth and production.** 2015. 78p. (Master's dissertation) - Institute of Agricultural and Technological Sciences, Federal University of Mato Grosso, Mato Grosso, 2015.
DUARTE, G. R. B.; SCHÖFFEL, E. R.; MENDEZ, M. E. G.; DE PAULA, V. A. Measurement and estimation of tomato evapotranspiration grown under organic fertilization in a protected environment. **Revista Semina**, Londrina, v. 31, n. 3, p. 563-574, 2010.
FANELA, T. L. M.; MOTA P. R. D; VILLAS BÖAS R. L.; LUDWIG F.; FERNANDES D. M. Influence of different levels of water tension on the gerbera crop grown in substrate. In: Amostra científica em ciências agrárias, 2.; mostra científica da fmvz, 10.; reuniao científica em ciências agrárias do lageado. Botucatu. **Anais**... Botucatu: UNESP, 2006.
FARIAS, M. F.; SAAD, J. C. C. Growth and quality of chrysanthemum grown in pots under a protected environment. **Horticultura Brasileira, Brasilia**, v. 23, n. 3, p. 740-742, jul-set, 2005.
FAVA, C. L. F., CAMILI, E. C. Production of *Anthurium andraeanum* cultivars in the conditions of Acorizal-Mt. **Horticultura Ornamental**, v. 20, n. 2, p. 179-184, 2014.
FERMINO, M. H. The use of physical analysis in assessing the quality of components

and substrates. In: FURLANI, A. M. C.; BATAGLIA, O. C.; ABREU, M. F.; ABREU, C. A.; FURLANI, P. R., QUAGGIO, J. A.; MINAMI, K. **Characterization, management and quality of substrates for plant production**. Campinas: Agronomic Institute, 2002. p.29-37.
FERMINO, M. H. **Substrates**: composition, characterization and methods of analysis. Guaíba: **Agro livros**, 2014. 112 p.
FERMINO, M.H. **Methods of analysis for the physical characterization of plant substrates**. 2003. 89p. Thesis (Doctorate in Plant Science). Federal University of Rio Grande do Sul. Porto Alegre, 2003.
FERREIRA, D. F. SISVAR: a computational system for analyzing statistics. **Ciência e agrotecnologia**, v. 35, n. 6, p. 1039-1042, 2011.
FRIZZONE, J. A.; ANDRADE JUNIOR, A. S. DE; SOUZA, J. L. M. de; ZOCOLER, J. L.; Irrigation planning. Investment decision analysis. 1. ed. **Brasília: Embrapa Informaçâo Tecnológica**, 627 p., 2005.
FRIZZONE, J. A. Bean plant response functions to the use of nitrogen and irrigation. In: SYMPOSIUM ON WATER MANAGEMENT IN AGRICULTURE, 1987, Campinas. **Proceedings**... Campinas: Unicamp, 1987. 123133 p.
FRIZZONE, J. A.; TEODORO, R. E. F.; PEREIRA A. S.; BOTREL, T. A.; Water table and nitrogen doses in the production of oats (*Avena sativa* L) for fodder. **Sci. Agric**. Piracicaba, v. 52, n.3, p. 578-586, 1995.
GIRARDI, L. B. **Water availability in the production of Alstroemeria (Alstroemeria x hybrida) in pots.** 2016. 90 p. Thesis (Doctorate in Agricultural Engineering) - Federal University of Santa Maria, Santa Maria. 2016.
GIRARDI, L. B.; PEITER, M. X.; BELLÉ, R. A.; BACKES, F. A. A.; SOARES, F. C.; VALMORBIDA, I. Water availability and its effects on root development and production of potted gypsophila in a protected environment. **Revista Irriga**, Botucatu, v. 17, n. p. 501 - 509, 2012.
GIRARDI, L. B.; PEITER, X. P.; BELLÉ, R. A.; ROBAINA A. D.; TORRES R. R.; KIRCHNER, J. H.; BEN, L. H B. Evapotranspiration and crop coefficient of alstroemeria (*Alstroemeria x hybrida*) grown in a greenhouse. **Revista Irriga**, Botucatu, v. 21, n. 4, p. 817-829, 2016.
GIRARDI, L.B.; PEITER, M. X. ; PIMENTA, B. D.; BRUNING, J.; RODRIGUES, S. A. R.; KIRCHNER, J. H. Growth and development of alstroemeria x hybrida when subjected to different pot retention capacities. **Revista Irriga**, Botucatu, v.11, n.1, p. 1191 - 1200, 2016.
GOMES, A. R. M.; D'Ávila, J. N. T.; GONDIM, R. S.; Bezerra, F. C.; BEZERRA, F. M. L. Estimation of evapotranspiration and crop coefficient of *Heliconia psittacorum* L. x H. spathocircinada (Arist) grown in a protected environment. **Revista Ciência Agronômica**, v.37, p.13-18, 2006.
GOMES, A. R. M.; GONDIM, R. S.; BEZERRA, F. C.; COSTA, C. A. G. Evapotranspiration and crop coefficients of *Alpinia purpurata*. **Revista Ciência Agronômica**, v.39, p.481-486, 2008.
GRUSZYNSKI, C. **Commercial chrysanthemum production: pot, cutting and garden**. Guaíba: Agropecuária, 166 p. 2001.
GUERRERO A.C.; FERNANDES D.M.; LUDWIG F. Nutrient accumulation in potted gerbera as a function of potassium sources and doses. **Horticultura Brasileira** 30: 201-208. 2012.
GUISELIN, C. **Microclimate and Production of Gerbera in Protected Environments with Different Types of Covering**.2002. 71 p. (Master's dissertation) Piracicaba. Sao Paulo. 2002.
GUISELINI, C.; SENTELHAS, P. C.; PANDORFI, H.; HOLCMAN, E. Cover management of protected environments: solar radiation and its effects on gerbera

production. **Revista Brasileira de Engenharia Agrícola e Ambiental**, v.14, p.645-652, 2010.
HEXEM, R. W.; HEADY, E. O. **Waterproductionfunction for irrigatedagriculture**. Ames: Iowa State University, 1978. 215 p.
HULSHOF, T. **Gérberas**: on nutritional disorders and fertilization in cut gerbera cultivation. Holambra: Editora Setembro. 2008. 48 p.
IBRAFLOR. Brazilian Institute of Floriculture. **Gerbera classification criteria**. Sao Paulo, 2017. Cooperativa Velling Holambra, Flores e Plantas Ornamentais. Available at http://www.ibraflor.com/p_qualidade.php. Accessed on: October 3, 2017.
IBRAFLOR. Brazilian Institute of Floriculture. **Domestic market data**. Sao Paulo, 2014. Available at: http://www.ibraflor.com/ns_mer_interno.php. Accessed on: Dec. 10, 2017.
IBRAFLOR. Brazilian Institute of Floriculture. **Sector figures: Domestic market**. Sao Paulo, 2017. Available at: <http://www.ibraflor.com/publicacoes/vw.php?cod=256>. Accessed on: 02 feb.2017.
INFOAGRO. **Gerbera cultivation**. 2012. Available at: <http://www.infoagro.com/flores/flores/gerbera.htm>. Accessed on September 19, 2017.
INFOAGRO. **Gerbera cultivation**. 2014. Available at: http://www.infoagro.com/flores/flores/gerbera.htm. Accessed on November 26, 2017.
JUNQUEIRA, A. H.; PEETZ, M. S. (2005) **Perfil da cadeia produtiva de flores e plantas ornamentáis da mesorregiao metropolitana de Belém- PA**. Belém: SEBRAE / PA. (SEBRAE Edition) 49p.
LIBARDI, P.L. **Dinâmica da água no solo.** 2. ed. Piracicaba, ESALQ, 509 p. 2000.
LIMA JÚNIOR, J. C.; NAKATANI, J. K.; NETO, L. C. M.; LIMA, L. A. C. de V. de.; KALAKI, F. B.; CAMARGO, R. B. de. **Mapping and quantifying the flower and ornamental plant chain in Brazil** [coordination and organization Marcos Fava Neves; Mairun Junqueira Alves Pinto]. 1.ed. Sao Paulo: OCESP, 2015. 132p.
LIN, W. C.; FRENCH, C. J. Effects of Supplementary Lighting and Soil Warning on Flowering of Three Gerbera Cultivars, **Hort Science**, 20: 271273, 1985.
LUDWIG, F. Potted *gerbera* (*Gerbera jamesonii* L.) cultivars under two levels of fertigation. 2007. 79 f. Dissertation (Master's degree in Agronomy/Horticulture) - Faculdade de Ciências Agronomías, Universidade Estadual Paulista, Botucatu, 2007. **Potted gerbera (*Gerbera jamesonii* L.) cultivars under two levels of fertigation**. 2007.
LUDWIG, F.; FERNANDES, D.M.; GUERRERO, A.C.; VILLAS BOAS, R.L. LASCHI, D. Substrates in the development of potted gerbera cultivars. **Revista Brasileira de Horticultura Ornamental**. v. 21, No.2, p. 177-184, 2015.
LUDWIG, F.; GUERRERO, A. C.; FERNANDES, R. L.; VILLAS BÔAS, R. L. (2010b) Growth analysis of potted gerbera grown in different substrates. **Horticultura Brasileira**, 28:70-74.
LUDWIG, F; FERNANDES, D. M.; MOTA, P. R. D.; VILLAS BOAS, R. L. Growth and production of gerbera fertirrigated with nutrient solution. **Horticultura Brasileira**, Brasilia, DF, v. 28, n. 4, p. 424429, Oct./Dec. 2010.
MALAVOLTA, E.; VITTI, G. C.; OLIVEIRA, S. A. de. **Evaluation of the nutritional status of plants - principles and applications**. 2 ed: Brazilian association for potash and phosphate research. 1997. 319p.
MARQUES SILVA, P. A.; PEREIRA MAGELA, G.; REIS, R.; LIMA, L. A.; TAVEIRA J. H.; Response function of American lettuce to water levels and nitrogen fertilization. **Ciênc. Agrotec.**, v. 32, n.4, p.1266-1271, 2008.
MASCARINI, L. Gerbera cultivation in growing media. **Horticulture**

International. Tarragona, v. 6, n. 19, p. 86-88, 1998.
MEDEIROS, S. de S. et al. Use of urban wastewater in the cultivation of gerberas: effects on production components. **Revista de Engenharia Agrícola**. Jaboticabal, v. 27, n. 2, p. 569-578, May/Aug. 2007.
MELLO, R. P. **Water consumption of the Asiatic Lily in pots with different substrates**. 2006. 74 p. Master's Degree Dissertation - Federal University of Santa Maria, Santa Maria, RS, 2006.
MENEGAES, J. F.; SWAROWSKY, A.; BACKES, F. A. A. L.; BELLÉ, R. A.; FILHO, H. J. I. Diagnosis of the retail flower market of Santa Maria, RS. **Ornamental Horticulture**, v. 21, n. 3, p. 291-298, 2015.
MENEGAES, J. F.; SWAROWSKY, A.; BACKES, F. A. A. L.; BELLÉ, R. A.; FILHO, H. J. I. Water consumption of calla lily submitted to soil irrigation management and copper content. **Revista Irriga**, Botucatu, v. 22, n. 1, p. 74-86, January-March, 2017.
MERCURIO, G. **Gerbera cultivation in greenhouse**. The Netherlands: Schreurs. 2002 206 p.
MONTEIRO, R. O. C.; COLARES, D. S.; COSTA T. R. N.; LEÂO, M. C. S.; DE AGUIAR, J. V.; Response function of melon to different irrigation rates and nitrogen doses. **Horticultura Brasileira**. v. 24, n. 4, p. 455459, 2006.
MONTEIRO, R. O. C.; NONATO TÁVORA R.; LEÂO SARAIVA C.; DE AGUIAR VANGLÉSIO J.; Aspectos econômicos da produçâo de melâo submetido a diferentes lminas de irrigaçâo e doses de nitrogênio. **Irriga**, Botucatu, v.12, n.3, p. 364-376, 2007.
NEVES, M. F.; PINTO, M. J. A. **Mapping and Quantifying Brazil's Flower and Ornamental Plant Chain**. Sâo Paulo: OCESP, 2015.
Neto, O. N. de S. **Moisture limits and potassium concentrations in the soil solution when growing lisianthus (*Eustoma grandiflorum*) fertirrigated in a protected environment**. 2017,113 p. PhD Thesis.
Piracicaba. University of Sao Paulo. Luiz de Queiroz College of Agriculture. Piracicaba. 2017.
NOYA, M. G.; CUQUEL, F. L.; ARMINDO, R. A.; SOUZA, J. L. M. Cultivation of *Stenachaenium megapotamicum* under different irrigation regimes. **Ciencia Rural**, Santa Maria, v. 44, n. 1, p. 79-84, 2014.
OLIVEIRA, E. C.; CARVALHO, J. A.; REZENDE, F. C.; ALMEIDA, E. F. A.; REIS, S. N.; MIMURA, S. N. Evapotranspiration of rose bushes grown in a protected environment. Revista Brasileira de Engenharia Agrícola e Ambiental, Campina Grande, v. 18, n. 3, p. 314-321, 2014.
OLIVEIRA, E. C.; CARVALHO, J. A.; REZENDE, F. C.; ALMEIDA, E. F. A.; REIS, S. N.; MIMURA, S. N. Yield of roses grown in a protected environment under different irrigation levels. **Revista Irriga**, Botucatu, v. 21, n. 1, p. 14-24, 2016.
OLIVEIRA, G.M; LEITÄO, M. M. B. R; BISPO, R.C; SANTOS, I. M. S; LIMA, C. B. A; CARVALHO, A. R. P. Crop coefficient and productivity of onion submitted to different irrigation laminas. **Brazilian Journal of Agricultural and Environmental Engineering**. Campiña Grande, PB v.17, n.9, p. 969974, 2013.
OLIVEIRA, S. L. **Response functions of sweet corn to the use of irrigation and nitrogen**.1993. 91 f. Thesis (Doctorate in Irrigation and Drainage)
Federal University of Viposa, Viposa, 1993.
PADRÓN, R. A. R.; NOGUEIRA, H. M. C. M.; CERQUERA, R. R.; ALBINO, G. D.; NOGUEIRA, C. U. Physical-water characterization of yellow clay loam soil for establishing irrigation design and management. **Acta Iguazu**, Cascavel, v.4, n.1, p. 36-47, 2015.
PANJ, F.G., KUMARI, S.; PARMAR, P. B. Effect of growing media properties and its correlation study in gerbera production. **The Bioscan**. v. 9, n. 1, p. 7983, 2014.

PATRA, S. K.; BEURA, S. Media standardization for pre-hardening and hardening of in vitro regenerated plantlets of gerbera cultivars. **Journal of Applied Horticulture**. v. 18, n 1, p. 76-79, 2016.
CASTRO, R. P.; COSTA, R. N. T.; SILVA, L. A. C.; GOMES FILHO, R. R. Decision models for economic optimization of water use in irrigated areas of the Vale do Curu Experimental Farm, Pentecoste - CE. **Irriga**, Botucatu, v. 12, n. 3, p. 377-392, 2007.
PENMAN, H. L. Natural evaporation from open water, bare soil and grass. **Proc. Royal Soc. A**, London. v 193, p. 120-143, 1948.
PEREIRA, A. R.; VILLA NOVA, N. A.; SEDIYAMA, G. C. **Evapo(transpi)raçâo**. Piracicaba: FEALQ. 183p. 1997.
PEREIRA, J. R. D; CARVALHO J.A; PAIVA P.D.O; SILVA, E.L; FAQUIN, V. Efeito da época de suspensão da fertirrigaçao e níveis de reposição de água na cultura do crisântemo (*Dendranthema grandiflora*). **Ciência e Agrotecnologia**. 27: 658-664, 2003.
PEREIRA, J. R. D. **Technical and economic viability of water and nitrogen applications in the cultivation of gladiolus (*Gladiolus x grandiflorus* L.).** 2005. 80 p. Doctoral thesis - Federal University of Lavras, Lavras, 2005.
PEREIRA, L. G. **Flower Stem Production in Gerberas Subjected to Different Soil Water Tensions**. 2013. Dissertation (Master's Degree in Water Resources in Agricultural Systems), Federal University of Lavras, Lavras, 2013.
PEREIRA, M. T. J.; SILVA, T. J. A.; BONFIN-SILVA, E. M.; MAZZINI- GUEDES, R. B. **Applying wood ash and soil moisture on gladiolus (*Gladiolus grandiflorus*) cultivation.** Australian Journal of Crop Science, v. 10, n. 3, p. 393-401, 2016.
PEREIRA, O. C. N.; BERTONHA, A.; FREITAS, P. S. L. DE; GONÇALVES, A. C. A.; REZENDE, R.; SILVA, F. A. da;. Lettuce production as a function of water and nitrogen. **Acta Scientiarum Agronomy**, Maringá, v. 25, n.2, p. 381387, 2003.
PIVETTA, C. R.; HELDWEIN, A. B.; MALDANER, I. C.; RADONS, S. Z.; TAZZO, I. F.; LUCAS, D. D. Maximum evapotranspiration of peppers grown in a plastic greenhouse as a function of phenometric and meteorological variables. **Revista Brasileira de Engenharia Agrícola e Ambiental**, v.14, p.768-775, 2010.
POLYSACK. Available at: http://www.polysack.com/ index.php?page_id=214. Accessed on 19/11/2017.
RAVIV, M.; BLOM, T. J. The effect of water availability and quality on photosynthesis and productivity of soilless-grown cut roses. **Scientia Horticulture**. Amsterdam, n.88, n.4, p.257-276, 2001.
REICHARDT, K.; TIM, L.C. **Soil, plant and atmosphere: concepts, processes and applications**. Sao Paulo: Manóle, 2004. 478p.
REIS, L. S.; SOUZA, J. L.; AZEVEDO, C. A. V. de. Evapotranspiration and crop coefficient of kaki tomato grown in a protected environment. **Revista Brasileira de Engenharia Agrícola e Ambiental**, v.13, p.289-296, 2009.
RIBEIRO, M. S.; SILVA, E. L.; MOURA, D. C. M.; DANTAS, A. A. Crop *coefficients* (kc) and vegetative growth of Acaiá Cerrado (*Coffea arabica* L.) associated with degrees of development. **Revista Irriga**, Botucatu, v. 14, p. 220-232, 2009.
ROGERS, M. N. and TJIA, B. O. S. 1990. **Gerbera production for cut flowers and pot plants**. Timber Press, Portland, Oregon. 116 p.
ROGERS, M. N., TJIA, B. O. Gerbera production. Timber Press Growers handbook series, Portland, Oregon. v.4, 116 p. 1990.
ROMAHN, V. **Illustrated encyclopedia 2200 plants and flowers. Flowers and foliage: Herbaceous plants.** Vol. 3, Editora Europa. 274 p. 2007.
SANTIAGO, E. J. P. et al. Adjustment of the Hargreaves and Samani equation from lysimeter data for the municipality of Juazeiro-BA.**Revista Irriga,** Botucatu, p. 108-114, 2016.

SAVVAS, D.; GIZAS, G. Response of hydroponically grown gerbera to nutrient solution recycling and different cation ratios. **Scientia Horticulturae**, Amsterdam, v. 96, p. 267-280, 2002.
SCHAFER, G.; SOUZA, P.V.D.; FIOR, C.S. An overview of the physical and chemical properties of substrates used in horticulture in southern Brazil. **Revista Brasileira de Horticultura Ornamental**, Campinas, v. 21, n.3, p. 299-306, 2015.
SCHAFER, G.; SOUZA, P.V.D.; KOLLER, O.C.; SCHWARZ, S.F. Physical and chemical properties of substrates to cultivate seedling of citrus rootstocks. **Communications in Soil Science and Plant Analysis**, New York, v. 39, p. 1067-1079, 2008.
SCHMITZ, J.A.K.; SOUZA, P.V.D.; KAMPF, A.N. Chemical and physical properties of mineral and organic substrates for growing seedlings in containers. **Ciencia Rural**, Santa Maria, v.32, n.6, p.937-944. 2002.
SCHWAB, N. T., PEITER, M. X., BELLÉ, R. A., BACKES, F. A. A., ROBAINA, A. D., FERRAZ, R. C. Water consumption of cravina submitted to different irrigation strategies and pot sizes. **Revista Irriga**, Botucatu, v. 18, n. 2, p. 328-336, 2013.
SCHWAB, N. T. Water availability in the cultivation of cravina in pots with rice husk ash substrate. 2011, 80 p. Dissertation (Master's in Agricultural Engineering) - Federal University of Santa Maria, Santa Maria, 2011.
SILVA, D. de O. **Growing gerbera under canopies and water stress in the São Francisco Valley**. 2017. Dissertation (Master's Degree in Irrigated Horticulture), Bahia State University, Bahia, 2017.
SILVA, M. T. Evapotranspiration and crop coefficients of irrigated peanut under edaphoclimatic conditions in the Cariri region of the State of Ceará. **Journal of Biology and Earth Sciences**. V.8, n.1, 2008.
SILVA, W. L. C.; MAROUELLI, W. A. Irrigation management for vegetables in the field and in protected environments. In: FARIA, M. A. (Coord.). **irrigation management**. Lavras: UFLA/SBEA, 1998. p. 311-348.
SOARES, F. B.; PEITER, M. X.; ROBAINA, A. D.; PARIZI, A. R. C.; RAMÂO, C. J.; VIVAN, G. A. Yield responses of maize hybrids grown under different irrigation strategies. **Revista Irriga**, Botucatu, v.15, n. 1, p.36-50, 2010.
SONNEVELD, C.; VOOGT, W. Effects of pH and Mn application on yield and nutrient absorption with rockwool grown gerbera (refereed). **Acta Horticulturae**, Wageningen, v.450, p.139-147, 1997.
SOUZA, V. C.; LORENZI, H. (2008). **Systematic botany. Illustrative guide to the identification of native and exotic phanerogamous families in Brazil**, based on APG II. 2nd edition. Instituto Plantarum. 704p.
TAHIR, M. H. N., IMRAN, M.; HUSSAIN, M. K. Evaluation of sunflower (Helianthus annuus L.) inbred lines for drought tolerance. **International Journal Agricultural Biology**. v. 1, n. 3, p. 398-400, 2002.
TARSITANO, M. A. A.; HOFFMAN, R. Análise económica do emprego de fertilizantes na cultura do milho. **Revista Economía Rural**, Sao Paulo, v. 23, n. 3, p. 333-349, 1985.
TAVARES, R. Constructing concept maps. **Ciências & Cogniçao**, Rio de Janeiro. v. 12, p. 72-85, Oct/Dec. 2007.Available at :< http://www.cienciasecognicao.org/pdf/v12/m347187.pdf>. Accessed on: October 5, 2017.
TERRA, S. B.; ZÜGE, D. P. P. D. O. Floricultura: a produção de flores como uma nova alternativa de emprego e renda para a comunidade de Bagé- RS. **Conexao UEPG magazine**. Ponta Grossa, v. 9, n. 2 - jul./dez. 2013.
TSIROGIANNIS, I.; S KATSOULAS, N.; KITTAS. C. 2010. Effect of Irrigation Scheduling on Gerbera Flower Yield and Quality. **HortScience**, v. 45, n. 2, p. 265-270,

2010.
VALERIANO, T. T. B.; SANTANA, M. J.; SOUZA, S. S. de.; PEREIRA, U. da.; CAMPOS, T. M. Optimum economic blade for irrigated tomato cv. andréa in a protected environment. **Revista Inova Ciência & Tecnologia**, Uberaba, p. 13-19, v. 3, n. 2, jul/dez., 2017.
VEATCH-BLOHM, M.E., MALINOWSKI, M., KEEFER, D. Leaf water status, osmotic adjustment and carbon assimilation in colored calla lilies in response to saline irrigation. **Scientia Horticulturae**, Amsterdam, v. 144, p. 65-73, 2012.
VEILING HOLAMBRA. RES: Veiling cooperative data Holambra.<http://www.veiling.com.br/>. Accessed on: 12/11/2017.
VERDONCK, O.; GABRIELS, R. Substrate requirements for plants. **Acta Horticulturae, Wageningen,** v. 221, p. 19-23, 1988.
VIDA, J.B.; KUROZAWA,C.; ESTRADA, K.R.F.S & SANTOS, H.S. **Plant health management in protected cultivation**. In: Goto, R & Tivelli, S.W. (Eds.) Vegetable production in protected environments: subtropical conditions. Botucatu. UNESP, 1998. pp 53-104.
Vilas-Boas, R. C.; Carvalho, J. de A.; Gomes, L. A. A.; Sousa, A. M. G de; Rodrigues, R. C.; Souza, K. D. de. Technical and economic evaluation of the production of two cultivars of curly lettuce as a function of irrigation rates. **Ciência e Agrotecnologia**, v.32. p.525-531, 2008.
ZAMBOLIM, L., COSTA, H., LOPES, C.A.; VALE, F. X. R. **Vegetable diseases in protected cultivation**. Informe Agropecuário 20: 114-125. 1999.
ZAMBOLIM, L.; COSTA, H.; LOPES, C.A & VALE, F.X.R. **Vegetable diseases in protected cultivation**. In: Zambolin, L.; Vale, F.X.R.; COSTA, H. (Eds.). Control of vegetable plant diseases. V1. Viposa. Federal University of Viposa, 2000. p. 373-407.
ZENG, C.; BIE, Z.; YUAN, B. Determination of optimum irrigation water amount for drip-irrigated muskmelon (*Cucumis melo* L.) in plastic greenhouse**. Agricultural Water Management**, Amsterdam, v. 96, n. 4, p. 595-602. 2009.
ZORZETO, T. Q.; DECHEN, S. C. F.; ABREU, M. F.; FERNANDES JÚNIOR, F. Physical characterization of substrates for plants. **Bragantia**, v.73, n.3, p.300-311, 2014.

ANNEXES

ANNEX A - Classification criteria for Cut Gerbera - VEILING HOLAMBRA.

GÉRBERA DE CORTE

Classificar é separar os produtos em lotes homogêneos quanto ao padrão e qualidade, caracterizados separadamente. O critério de classificação é o instrumento que unifica a comunicação entre toda a cadeia de produção. Produtores, atacadistas, varejistas, consumidores precisam seguir os mesmos critérios para determinar a qualidade do produto. Assim, haverá mais transparência na comercialização, valorização do melhor produto, maior qualidade e maior consumo.

PADRÃO. São as características mensuráveis do produto. O Padrão é determinado pela uniformidade do lote. O lote de Gérbera padronizado é aquele que possui **95% de uniformidade** quanto ao comprimento, a espessura e o tamanho da flor.

Comprimento da haste

É determinado pelo tamanho da haste desde a sua ponta até o miolo da flor, obedecendo à tabela abaixo.

Tamanho da Flor

A caixa de Gérbera deverá ser classificada conforme os tamanhos da flor **em >10 ou <10 cm.**
Admite-se uma variação dentro da mesma caixa entre a maior e a menor flor de até 1,5 cm para cima do tamanho discriminado na caixa.

Código do Produto	Comprimento da Haste	Tamanho da Flor	Descrição do Produto	Descrição dos Códigos
02027.050.000.10.01	**050**	**10**	**GERB. ORCA>10**	**HASTE LONGA/FLOR GRANDE**
02027.050.000.71.01	**050**	**71**	**GERB. ORCA<10**	**HASTE LONGA/FLOR PEQUENA**
02027.250.000.10.01	**250**	**10**	**GERB. ORCA>10**	**HASTE CURTA/FLOR GRANDE**
02027.250.000.71.01	**250**	**71**	**GERB. ORCA<10**	**HASTE CURTA/FLOR PEQUENA**

Legenda:

- 050 (= ou > que 0,50 cm)
- 250 (de 35 a 49 cm)
- 71 (diâmetro da flor < que 10 cm)
- 10 (diâmetro da flor > que 10 cm)

Departamento Qualidade
1

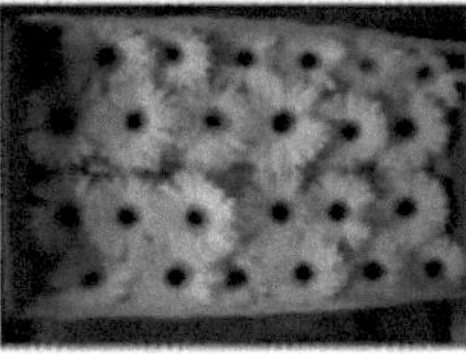

Ponto de abertura irregular. Pontos de aberturas diferentes dentro da mesma caixa. Admite-se até 5 flores com 50% do miolo aberto.

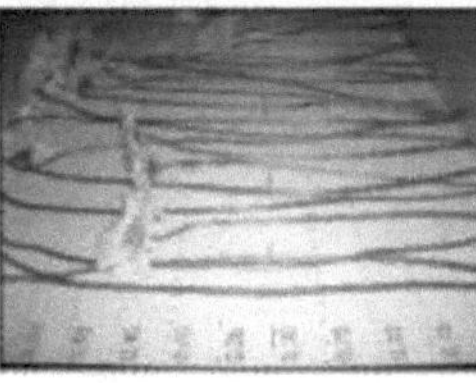

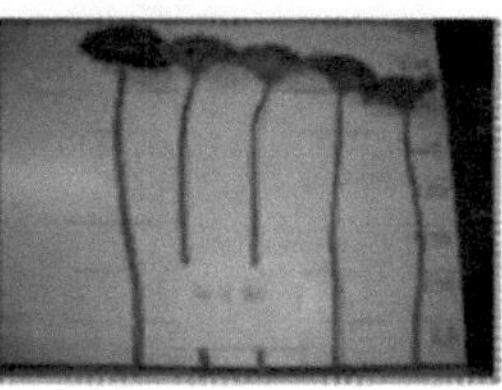

Espessura da haste

A seleção do lote por espessura serve para dar uniformidade ao lote.

A haste de Gérbera deve estar suficientemente forte e hidratada para sustentar o peso da flor. Com base no comprimento, as hastes devem apresentar uniformidade de espessura, possuindo entre 5 e 6 mm.

OBS: As hastes devem ser firmes e com sustentação sem desvios da flor acentuados.

QUALIDADE. É a ausência de defeitos.

A categoria de classificação (A1 ou A2) caracteriza a qualidade do lote e deverá ser estabelecido conforme limites de tolerâncias para defeitos graves e leves. Os defeitos graves são aqueles que podem continuar a evoluir durante o processo de comercialização. O produtor deverá selecionar o produto, eliminando os defeitos antes do embalamento, assegurando requisitos mínimos de qualidade, abaixo do qual o produto não poderá ser comercializado.

Defeitos
(Hastes na Caixa)

Defeitos Graves	A1	A2
Desidratação	0	0
Danos de botrytis	0	0
Danos de bolor no miolo (sclerotinia)	0	0
Dano de míldio ou oídio	Até 3	Até 5
Dano de pragas (trips, ácaros, pulgão)	0	Até 3
Danos mecânicos	Até 3	Até 5
Defeitos Leves	**A1**	**A2**
Deformação da haste	Até 5	Até 8

Tab.- Tabela para determinação da categoria de qualidade de acordo com a tolerância aos defeitos.

Defeitos Graves

São aqueles que depreciam a aparência e desvalorizam a qualidade do produto, podendo aumentar de intensidade com o tempo, restringindo ou inviabilizando sua comercialização.

Desidratação. Perda de água dos tecidos visualizados por evidente murcha da flor ou haste. O produto que apresentar início de murchamento será devolvido ao produtor;

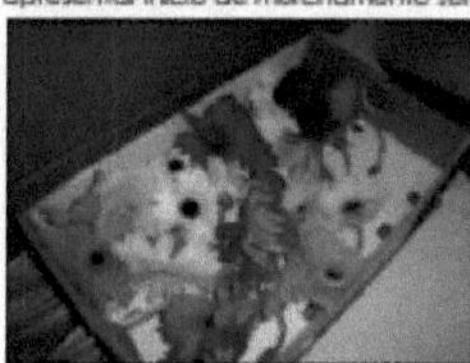

Danos de botrytis. Danos de apodrecimento do tecido da flor se iniciam com pequenas pintas nas pétalas. Na presença do dano visível, o produto será desclassificado para "B" automaticamente, dependendo da sua intensidade;

Danos de bolor no miolo (Sclerotinia). Presença do fungo esporulando no miolo da flor, sem tolerância para as ambas às categorias (A1 e A2);

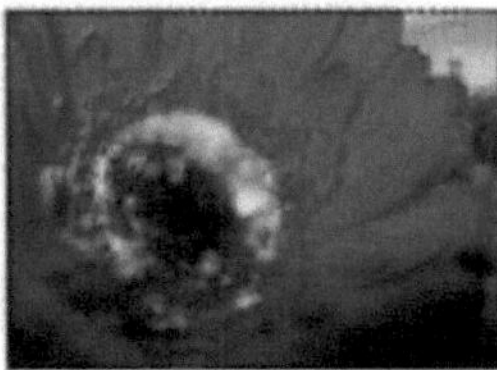

Danos de míldio ou oídio. Danos de diferentes características causadas pela infecção de agentes patogênicos;

Danos de pragas. Danos de diferentes características causadas pela infestação de insetos, ácaros e outros. O produto será submetido à inspeção para detectar a presença do inseto/ácaro controlado, mesmo que o sintoma não seja visível;

Danos mecânicos na flor. Danos causados pelo rompimento ou deformação superficial do tecido da flor provocada por ação mecânica. Admite-se até 3 flores com danos mecânicos leves nas pétalas das flores da borda da caixa. Produtos com hastes quebradas serão devolvidos.

Defeitos Leves

São aqueles que depreciam a qualidade, mas não evoluem com o tempo causando mudança na aparência até o destino final.

Deformação da haste. Desvio da forma retilínea da haste. Não deverão ser enviados produtos com haste apresentando formato de **"S"** ou curvatura maior que 10 cm em relação à linha horizontal.

INFORMAÇÕES ADICIONAIS.

Tratamento pós-colheita com solução conservante é obrigatório;

O produto deve estar previamente hidratado para evitar murchamento e assegurar durabilidade na comercialização;

A caixa deve conter a informação do comprimento e diâmetro da flor obrigatoriamente;

Nas caixas de gérberas será aceito até 4 hastes fora do padrão que foi colocado na boleta e na caixa do produto. Caso o número de hastes seja superior a 4 hastes o produto será desclassificado para A2;

Tempo estocagem máximo de 03 dias de câmara fria;

O produto deverá chegar previamente refrigerado do Sítio.

COOPERATIVA VEILING HOLAMBRA

Annex B - Potted gerbera classification criteria - VEILING HOLAMBRA

GÉRBERA DE VASO

Classificar é separar os produtos em lotes homogêneos quanto ao padrão e qualidade, caracterizados separadamente. O critério de classificação é o instrumento que unifica a comunicação entre toda a cadeia de produção. Produtores, atacadistas, varejistas, consumidores precisam seguir os mesmos critérios para determinar a qualidade do produto. Assim, haverá mais transparência na comercialização, valorização do melhor produto, maior qualidade e maior consumo.

PADRÃO. São as características mensuráveis do produto. O Padrão é determinado pela uniformidade do lote. O lote classificado da Gérbera Pote 12 deverá apresentar **90% de uniformidade** quanto à altura, nº flores e ponto de abertura.

Altura da planta

É determinado pelo tamanho da planta desde a borda do vaso até a média final das hastes florais, medido do centro do vaso.

Tamanho do Vaso	Altura da Planta Mínima	Altura da Planta Máxima
Pote 12 (midi)	**12 cm**	**25 cm**

Para a montagem do lote, recomenda-se ao produtor uma diferença de **até 5,0 cm entre o vaso mais alto e o mais baixo** para que não haja desuniformidade do lote quanto à altura.

OBS: Produtos abaixo de 12 cm e acima de 26 cm serão desclassificados para A2.

Planta abaixo da altura mínima

Planta acima da altura máxima

Departamento Qualidade
1

Formação da planta

Refere-se ao aspecto e constituição da planta.
O vaso deverá apresentar: Flores Abertas, folhas cobrindo totalmente o vaso, formação circular.

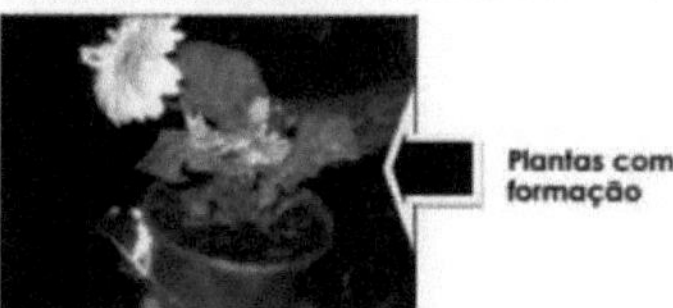

Quantidade de flores/vaso

Refere-se à quantidade mínima de flores/botões que o vaso deve apresentar no momento da comercialização.
O Vaso de Gérbera Pote 12 deverá ter no **mínimo 02 hastes florais abertas e bem formadas** para a comercialização.

Ponto de Abertura

Refere-se ao ponto de maturação no qual o produto é comercializado.
Excesso de maturação é a haste floral que apresenta um avançado estágio de maturação ou envelhecimento.

QUALIDADE. É a ausência de defeitos.
A categoria caracteriza a qualidade do lote, estabelecendo tolerâncias diferentes para os defeitos graves e leves. O produtor deverá eliminar os defeitos graves antes do embalamento, entretanto alterações que podem ocorrer no produto durante o processo de comercialização exigem o estabelecimento de tolerâncias aos defeitos graves que poderá se desenvolver durante o transporte ou depois que o produto já estiver nas mãos de clientes.

Departamento Qualidade
2

Defeitos
(Vasos no Carrinho)

Defeitos Graves	A1	A2
Desidratação	0	0
Danos de botrytis	0	0
Danos de bolor no miolo (sclerotinia)	0	0
Danos de pragas (trips, ácaros, lagartas, bicho mineiro)	8	16
Flor dupla	8	16
Danos mecânicos na flor	0	0
Danos de míldio ou oídio	0	8
Defeitos Leves	**A1**	**A2**
Queima por fitotoxidez	0	0
Resíduo químico (baixa intensidade)	8	16
Deficiência (ou falta) nutricional	0	0

Tab.- Tabela para determinação da categoria de qualidade de acordo com a tolerância aos defeitos.

Defeitos Graves

São aqueles que depreciam a aparência e desvalorizam a qualidade do produto, podendo aumentar de intensidade com o tempo, restringindo ou inviabilizando sua comercialização.

Danos de botrytis. Danos de apodrecimento do tecido da flor, sem tolerância para qualquer estágio de desenvolvimento da doença;

Departamento Qualidade
3

Danos de míldio, oídio ou fungos. Danos de diferentes características causadas pela infecção de agentes patogênicos;

Desidratação. Perda de água dos tecidos visualizados por evidente murcha da planta;

Danos de bolor no miolo (sclerotinia). Visualizado pela formação de fungos no miolo da flor;

Danos de pragas. Danos de diferentes características causadas pela infestação de insetos, ácaros e outros. Admite-se a presença dos danos apenas nas folhas da camada inferior da planta;

Danos mecânicos. Danos causados pelo rompimento ou deformação superficial do tecido da flor/ folha provocada por ação mecânica.

Defeitos Leves

São aqueles que depreciam a qualidade, mas não evoluem com o tempo causando mudança na aparência até o destino final.

Queima por fitotoxidez. Mancha de diferentes características decorrentes da toxidez, geralmente apresenta aspecto de queima nas bordas;

Resíduo químico. Manchas difusas leves que recobrem a epiderme dando um aspecto esbranquiçado à planta;

Deficiência nutricional. Desnutrição da planta visualizada por amarelecimento das folhas.

INFORMAÇÕES ADICIONAIS.

Deixar um espaço de no mínimo de 10 cm entre divisória do carrinho e a ponta da flor, para evitar danos mecânicos;

É obrigatório fazer a limpeza dos vasos antes de serem enviados para a comercialização;

Todos os vasos comercializados deverão vir devidamente preenchidos com substrato;

Para os produtos decorados o produto enviado deverá ser A1.

COOPERATIVA VEILING HOLAMBRA

Departamento Qualidade
6

Annex C - General report on the analysis of the water peat substrate for cut *gerbera (Gerbera jamesonii)*, carried out at the Biotechnology Laboratory - Analysis of Horticultural Substrates - Federal University of Rio Grande do Sul - UFRGS. Santa Maria - RS, 2017.

A M	EC	pH	DU	DS	U A	P T	E A	A D	A F D	A T	A R	CR A1 0	C R A 5 0	CR A 100

	mS .c m-1	(H 2O)	Kg m-3	Kg m-3	%	%	%	%	%	%	%	%	%	%
1	0,7 9	6, 74	57 0,1 6	25 6,9 8	54 ,9 3	89 ,0 5	15 ,2 2	35 ,1 0	31 ,3 0	3, 8 0	38 ,7 3	73, 83	4 2, 5 3	38, 73
2	0,8 0	6, 62	57 5,1 5	25 5,7 3	55 ,5 4	88 ,4 1	16 ,2 3	32 ,8 3	30 ,1 4	2, 6 9	39 ,3 5	72, 18	4 2, 0 4	39, 35
3	0,7 7	6, 60	57 6,0 0	25 9,3 7	54 ,9 7	87 ,9 2	16 ,6 3	33 ,4 8	29 ,3 6	4, 1 2	37 ,8 1	71, 29	4 1, 9 3	37, 81

AM = samples; DU = wet density; DS = dry density; UA = actual moisture;
pH = determined in water, diluted 1:5 (v/v); EC = electrical conductivity obtained in a 1:5 (v/v) solution.
PT = total porosity; EA = aeration space; AFD = easily available water; AT = buffering water; AD = available water; AR = remaining water
CRA10 = water retention capacity under suction of 10 cm water column; CRA50 = water retention capacity under suction of 50 cm water column; CRA100 = water retention capacity under suction of 100 cm water column.

Annex D - Report on the analysis of the water peat substrate used to grow cut gerbera *(Gerberajamesonii)*, carried out at the Biotechnology Laboratory - Analysis of Horticultural Substrates - Federal University of Rio Grandedo Sul - UFRGS. Santa Maria - RS, 2017.

Indicators	Substrate values
pH ($_{H2O}$)	6,50
EC (mS cm $)^{-1}$	0,78
DU (kg m $)^{-3}$	573,77
DS (kg m $)^{-3}$	257,36
UA (%)	55,15
PT (%)	88,46
AS (%)	16,03
AFD (%)	30,27

AD (%)	33,81
AT (%)	3,54
RA (%)	38,63
CRA 10 (%)	72,44
CRA 50 (%)	42,17
CRA 100 (%)	38,63

DU = wet density; DS = dry density; UA = actual humidity. pH = determined in water, diluted 1:5 (v/v); EC = electrical conductivity obtained in a 1:5 (v/v) solution PT = total porosity; EA = aeration space; AFD = easily available water; AT = buffering water; AD = available water; AR = Remaining water. CRA10 = water retention capacity under suction of 10 cm of water column; CRA50 = water retention capacity under suction of 50 cm of water column; CRA100 = water retention capacity under suction of 100 cm of water column.
*average of three samples

Annex E - Difference between varieties of cut gerbera - VEILING HOLAMBRA - Category: Cut Flowers.

SCIENTIFIC NAME: ***Gerbera jamesonii***
POPULAR NAME: Cariba gerbera

SCIENTIFIC NAME: ***Gerbera jamesonii***
POPULAR NAME: Gerbera Zingaro

SCIENTIFIC NAME: ***Gerbera jamesonii***
POPULAR NAME: Forza gerbera

SCIENTIFIC NAME: ***Gerbera jamesonii***
POPULAR NAME: Aztec gerbera

SCIENTIFIC NAME: ***Gerbera jamesonii***
POPULAR NAME: Barbasona gerbera

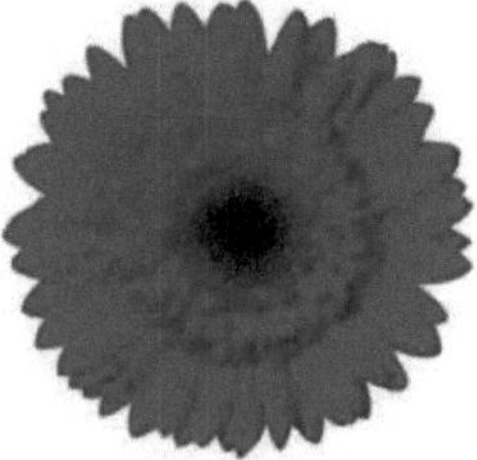

SCIENTIFIC NAME: ***Gerbera jamesonii***
POPULAR NAME: Beaudine gerbera

SCIENTIFIC NAME: ***Gerbera jamesonii***
POPULAR NAME: Explorer gerbera

SCIENTIFIC NAME: ***Gerbera jamesonii***
POPULAR NAME: Kendra gerbera

SCIENTIFIC NAME: ***Gerbera jamesonii***
POPULAR NAME: Lamborguini gerbera

SCIENTIFIC NAME: ***Gerberajamesonii***
POPULAR NAME: Special gerbera

SCIENTIFIC NAME: ***Gerbera jamesonii***
POPULAR NAME: Mixed Gerbera

SCIENTIFIC NAME: ***Gerbera jamesonii***
POPULAR NAME: Variegated Gerbera

Annex F - Quality standard for cut gerbera according to IBRAFLOR.

Parameters	Quality classes		
	Extra (A1)	Class I (A2J	Class II (B)

Phytosanitary aspects ¡taños	Stems and flowers free of pests, diseases and their damage, especially damage to the flowers, which must be free of *Botrytis* symptoms.	Stems and/or flowers with symptoms of pests and diseases that do not compromise the durability and appearance of the product.	Stems and/or flowers, with symptoms of pests and diseases, but which allow the products to be marketed.
Flowers	Flowers opening evenly in the box or wizard. Two rings of male flowers open in the center of the inflorescence. Intense coloring and healthy appearance.	Slightly uneven flower opening. A maximum of three rings of male flowers open on the inflorescence. Healthy color and appearance.	Uneven opening. Blemishes on the flowers that do not prevent the products from being marketed. Deformed flowers. More than three rings of open male flowers.
Presentation	Fresh products and excellent presentation.	Fresh and well presented products.	Products stocked and regularly presented.
Rods	Firm, straight, all the stems in the box of the same size (no broken stems). Minimum size: 40 cm.	Firm, straight, of the same size, at least 30 cm.	The stems are crooked, broken and uneven in size. Regular presentation, but in a condition to be marketed
Post-harvest treatment	Required.	Required.	Not done.

APPENDICES

Appendix A - Maximum air temperature (Tmax), minimum air temperature (Tmin) and relative humidity (RH) data for the cut gerbera growing season. Santa Maria - RS, 2017.

Date	Tmax (°C)*	Min. temperature (°C)*	RH (%)*
10/11/2016	35,1	16	65
17/11/2016	32,03	18,46	69,92
24/11/2016	35,6	18,37	63,83
01/12/2016	37,99	19,9	67,85
08/12/2016	37,13	18,73	61,92
15/12/2016	32,86	19,39	69,35
22/12/2016	34,54	19,44	66,93
29/12/2016	37,09	20,87	64,57
05/01/2017	37,86	20,96	68,15
12/01/2017	33,71	21,73	69,57
19/01/2017	37,21	22,11	72,07
26/01/2017	36,83	23,19	68,85
02/02/2017	40,27	22,97	72,57
09/02/2017	43,97	23,56	66,14
16/02/2017	37,91	20,61	69
23/02/2017	33,07	21,17	66,5
02/03/2017	32,66	18,97	74,28
09/03/2017	33,97	18,07	70,28
16/03/2017	34,3	20,3	65,57
23/03/2017	39,83	17,27	70,93
30/03/2017	40,59	20,71	67,35
06/04/2017	43,83	19,84	73,21
13/04/2017	38,27	14,89	73,64
20/04/2017	38,38	14,41	62,64
27/04/2017	29,7	15,79	67,14

1Data obtained inside the greenhouse of the Floriculture Sector of the Polytechnic College of the Federal University of Santa Maria - UFSM, 2016/2017.

Printed by Books on Demand GmbH, Norderstedt / Germany